Studies in National Governance and Emerging Technologies

Series Editors
Alberto Asquer
School of Finance
University of London, SOAS
London, UK

Edward Abbott-Halpin
Leeds Beckett University
Leeds, UK

More information about this series at
http://www.palgrave.com/gp/series/16181

Maxat Kassen

Open Data Governance and Its Actors

Theory and Practice

palgrave
macmillan

Maxat Kassen
Astana, Kazakhstan

ISSN 2524-6291 ISSN 2524-6305 (electronic)
Studies in National Governance and Emerging Technologies
ISBN 978-3-030-92064-7 ISBN 978-3-030-92065-4 (eBook)
https://doi.org/10.1007/978-3-030-92065-4

This Palgrave Macmillan imprint is published by the registered company Springer Nature Switzerland AG.
The registered company address is: Gewerbestrasse 11, 6330 Cham, Switzerland

Preface

The main idea of the book is to match both theoretical and practical knowledge about key actors and driving forces that help to initiate and advance such an emerging political and socioeconomic phenomenon as open data governance, resorting to the contexts of two Nordic nations, namely Finland and Sweden. For the last decade these two countries have been actively pioneering and advocating the concept of open data as one of the key points of reference to promote transparent and collaborative government in the context of an egalitarian society built on principles of social democracy, political representation and social justice among the majority of its members. In this regard, the results of the research presented in this work aim to shed light on the roles of key actors in the open data movement, which would help researchers to understand the key operational elements of data-driven governance by studying the most salient manifestations of related networking activities, motivations of stakeholders as well as political and socioeconomic readiness of public, private and civic sectors to advance such policies in their natural settings. Both driving forces and barriers were analyzed in the context of two countries, which are globally recognized as leaders of the open data movement, to propose a set of policy recommendations for practitioners in both public and private sectors of the economy. The book is intended for a global community of e-government experts and policymakers, political scientists, specialists of public administration, data journalists, lawyers and all who are interested in studying the truly interesting and promising phenomena of open data governance. Academically, the monograph could be also used by university faculty members and post-graduate students as a text for collateral reading in courses of e-government studies, public administration, public policy, political science, political economy, journalism and media communication, science and technology, democracy and regional studies.

The open data movement is beginning to play an ever-increasing role in public sector reforms all over the world. Likewise, the research on open data governance as a theoretical concept is on the rise in many academic disciplines and professional fields. However, related literature rarely focuses on the analysis of matching both theoretical knowledge and empirical evidence from its key actors and even less from multidimensional, cross-institutional and cross-border perspectives on how the movement is initiated and what its key operational elements, institutional and stakeholder roles are. In this connection, the main purpose of the book is to understand the dynamic and truly multidimensional role of the open data phenomenon that networks various political, socioeconomic and technological players around it in the context of two post-industrial and highly developed economies that for the last decade have been leading the open data movement in a global sense. The results of the research reveal an important role of political and socioeconomic communication among its key stakeholders, whether they are government agencies, businesses, citizens, developers, intermediaries, non-governmental organizations, academic or journalistic communities, by resorting to the study of a number of interesting cases in Finland and Sweden, which have recently adopted various collaborative and networking initiatives in the area. The main findings of the research suggest that the active promotion of the concept in private, non-governmental and civic sectors leads to the creation of new market niches where a promising phenomenon of open data-driven entrepreneurial and philanthropic activity of middlemen and intermediaries emerges, which is able to dramatically change traditional channels of public sector reforms and generation of related innovations in government.

Open data governance is an emerging phenomenon. In this regard, the results of the research are mostly intended as an introduction to this new concept from both theoretical and practical perspectives with many case studies and from various institutional contexts, proposing a wide array of practical recommendations for policymakers and open data entrepreneurs. Therefore, the book, which presents its content in a graphically illustrative manner with a lot of demonstrative cases and examples, could also be used as a reading for various academic programs and will be especially ideal for various professional courses and practical training for open data policymakers and practitioners. Moreover, considering that the monograph seeks to propose a universal methodological framework for related cross-country case studies that could be consistently applied to investigate the phenomena of open data governance in the context of many nations, especially in understanding key stakeholders of the open data movement and identifying its main political and socioeconomic drivers and challenges, the results of the research could also be used in evaluating and improving various strategic policies and executive decisions in the field. In this respect, the book is presented both as a policy review and as an agenda-setting research, which is intended for all those who are interested in understanding and studying the phenomenon from the perspective of its development in different political, economic, social and institutional contexts as well as from a wide array of theoretical and practical viewpoints.

In 2021 the world celebrated the 255th year of the first freedom of information law, which in the history of humankind was first adopted in both Finland and Sweden, when they were in a political union. In this connection, this work presents the case study research that investigates open data governance in the context of countries that are universally well-known today as pioneers of the global open knowledge movement that explicitly propagates the public values of free access to public information, which is mirrored in the widespread diffusion of other related promising concepts such as open science, open state, open source software, net neutrality and even pirate politics. In this regard, the main purpose of the book, which is generally based on the analysis of an extensive amount of empirical data that was obtained from the close observation during field studies in the Nordic region, document analysis of public reports, a number of semi-structured interviews and focus group discussions with various open data actors and online content analysis of related public and private collaborative platforms implemented in these two countries, is to explore an interesting world of open data governance from both stakeholder and institutional perspectives.

Resorting to the content analysis of a number of illustrative cases from Finland and Sweden that have recently adopted various comprehensive implementation strategies and networking initiatives in the area, the research also seeks to understand the dynamic and collaborative nature of related political and socioeconomic activities among its key stakeholders and locate typical drivers and challenges associated with the active promotion of the concept in both public and private sectors of economy. The case study of the phenomena in these two cultural contexts seems especially appealing, taking into account the recent rapid promotion of these countries in various global open data and e-government ratings. In this respect, one of the key findings of the research suggests that the successful adoption of the open data concept demands not only a fairly equal contribution from representatives of both public and private sectors of economy but, more importantly, new multi-level channels of political communication that could help them to network and collaborate with each other in advancing further independent open data-driven products and services.

Another finding of the study is that the active promotion of the concept among developers, non-governmental organizations and open data enthusiasts leads to the creation of new market niches, which has a profound multiplicative effect on the overall development of the concept as a multidimensional cross-institutional phenomenon. In this regard, all cases demonstrated that the true driver of the open data movement is a networking activity among both public and private players. Open data-driven public sector reforms could be successfully realized in a sustainable way only if various stakeholders, regardless of whether they are government agencies, citizens, independent developers, businesses, charities, think tanks or journalists, are politically and socioeconomically motivated to join the movement and contribute to the promotion of various participatory projects and civic engagement startups in the most collaborative manner.

Astana, Kazakhstan Maxat Kassen

Contents

About the Author

Maxat Kassen is a political scientist and e-government scholar from Kazakhstan. He is a former Fulbright Scholar in the University of Illinois at Chicago (USA). His research chiefly focuses on e-government and open data studies. He has published more than 50 academic articles on related topics, including in various international peer-reviewed impact factor journals and the author of three books, titled "Understanding systems of e-government" (Rowman & Littlefield: New York, 2015), "E-government in Kazakhstan: a case study of multidimensional phenomena" (Routledge: London, 2016) and "Open data politics" (Springer: Switzerland, 2019). In December 2017, Maxat Kassen was awarded with A.Baitursynov Silver Medal for significant achievements in scientific research and excellence in teaching students (The Association of Institutions of Higher Education, Almaty, Kazakhstan) and in November 2018 he received a prestigious international Scopus Award 2018 for Contribution to Science (Elsevier, The Netherlands). In addition, in November 2020 he won one more prestigious international academic distinction: The Web of Science – The Leader of Science Award 2020 in the category "Publication leader in social sciences" for his achievements in scientific research from such a global academic agency as Clarivate Analytics (Philadelphia, USA).

Abbreviations

CDO	Chief Data Officer
CKAN	Comprehensive Knowledge Archive Network
EC	European Commission
ERDF	European Regional Development Fund
EU	European Union
FOIA	Freedom of Information Act
GDP	Gross Domestic Product
ICTs	Information and Communication Technologies
IT	Information Technology
NGO	Non-governmental Organizations
ODC	Open Data Community
OGP	Open Government Partnership
OSCE	Organization for Security and Cooperation in Europe
OSF	Open Society Foundation
R&D	Research and Development
RTE	Real Time Economy
UN	United Nations
UNPAN	United Nations Public Administration Network
WJP	World Justice Project

List of Figures

List of Tables

CHAPTER 1

Open Data Governance as a Theoretical Concept: A Stakeholder and Institutional Analysis

Abstract The primary purpose of the study in this chapter, which methodologically relies on stakeholder and institutional analyses, is to comprehend theoretically the concept of open data through its understanding as a multidimensional and multi-institutional networking phenomenon. The diffusion of this intrinsically political rather than technocratic idea all over the world affects communication processes in many sectors of public life, providing new conceptual platforms to enhance civic engagement, direct participation and collaboration of various actors, which operate in the e-government area and transform traditional channels of political networking via new information technologies not only in interactions between main stakeholders of digital communication such as public agents, citizens and businesses but also between such new active players in the area as non-governmental organizations, independent developers, intermediaries and even mass media and, more importantly, between peers themselves in both public and private sectors of economy and, correspondingly, equally at global, national and local institutional levels of governance.

Keywords Open data • Open government • Structuration theory • Institutional theory • Stakeholder theory

M. Kassen, *Open Data Governance and Its Actors*, Studies in National Governance and Emerging Technologies,
https://doi.org/10.1007/978-3-030-92065-4_1

1 Introduction: Open Data Governance as a Multidimensional Networking Phenomenon

The first question that one could ask, starting reading this book, would be what open data is really as a phenomenon, how it is connected with government, what its manifestations are, who its actors are and why it is important to study all of these concepts under the umbrella of governance. It is especially essential to understand the meaning of the benevolent idea behind this promising phenomenon since from a semantic point of view, the scientific and media coverage of such a popular related concept as big data, which could be generated by a myriad of digital devices and sensors out there in private sectors of economy, occupies a lion's share of public and academic attention. Both concepts are indeed related in a sense that they are created automatically in a machine-to-machine readable manner by the mere existence and operation of the sophisticated digital infrastructure that human beings have managed to build in order to meet their needs in various niches of economy, in both public (e.g. government big data) and private sectors (corporate and civic big data). So the generation of raw data is a certain unintended positive side effect of the infrastructural operation in a typical developed nation. For example, the fact that data is collected in huge volumes helps many private corporations and sometimes government agencies to improve their business and operation models (Charalabidis et al., 2018; Kitchin, 2014) and meet better the needs of their customers in a highly tailored manner. Furthermore, some of these automatically collected public datasets could be then selected and published as open data in a transparent manner, of course, at the discretion of e-government policymakers, which could be reused and processed then by third parties to create new public services (see Fig. 1.1).

In this regard, the public value of open data is that it could potentially help to engage representatives of both non-governmental and private sectors of economy to join the movement and contribute in a meaningful manner by creating new open data-driven products and services or contributing to their initiation and promotion, which could be later consumed by all actors: citizens, businesses, academia, journalistic communities and, finally, the government itself, thus closing the loop. Indeed, being a provider of open data, the government initiates the distribution of public information, which not necessarily could be a final service in itself but rather a raw dataset, unprocessed and machine readable only. Such datasets could be generated within the existing infrastructural ecosystem, which

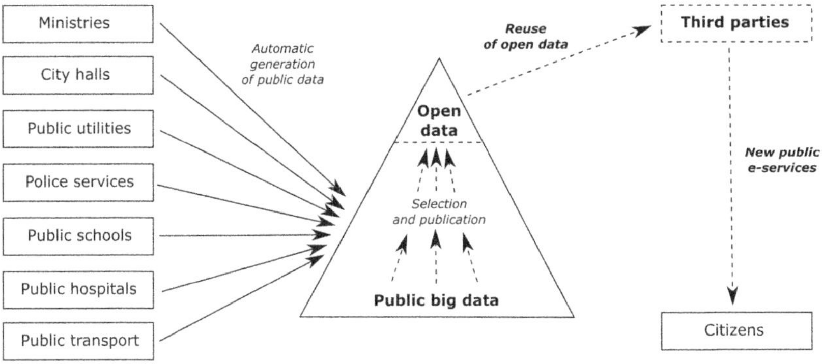

Fig. 1.1 Understanding the phenomena of open data. *Source: own illustration*

is already digitized and, therefore, easily monitored. Every transaction, decision-making, networking, event, operation and other activity of government bodies at all its institutional levels, regardless of whether they are at national, provincial and local levels, leave permanent digital blueprints, which could provide an invaluable source of information if they are published as open data, properly processed and then analyzed. Furthermore, this processing and analyzing could be done not necessarily by the public sector itself. As such, realizing the huge potential of the open data, governments around the world are starting to reveal the datasets about certain aspects of their own activities on a regular basis in raw machine-readable formats, hoping to initiate certain feedback, reaction and, more importantly, contribution generally from the society in further processing and turning these raw datasets into something that could be easily comprehensible and useful. Thus, open data is paradoxically becoming a certain antipode of big data since its value lies in the fact that it could be potentially reused by everyone and not only by those who have a government or corporate privilege to access it.

2 Understanding the Academic Background

In academic and professional literature, the open data governance phenomenon is usually analyzed in political and socioeconomic contexts of promising interactive communication processes that emerge between three traditional e-government stakeholders, namely public agents, developers and citizens (Kassen, 2017a; Lindman, 2014; Martin et al., 2011; Susha et al., 2015) and rarely in the background activity of such no less important players as businesses, academia, non-governmental organizations and journalists (Gonzalez-Zapata & Heeks, 2015; Safarov et al., 2017). Furthermore, most of the research in the field is concentrated on studying the attitude of particular stakeholders toward certain political, social, economic or technological manifestations of open data-driven public sector reforms such as the adoption policies (Kassen, 2018), life cycles of open data-driven innovations (Charalabidis et al., 2018), public values (Hjalmarsson et al., 2015) and usability of open data portals (Ojo et al., 2016) (see Table 1.1).

There is no academic work that would analyze the viewpoints of all related actors together in a more holistic manner and, more importantly,

Table 1.1 Identifying key researches on stakeholder perspectives about the open data movement

# *Authors (year) of publication*	*Stakeholder perspectives*	*Focus*
Martin et al. (2011)	Citizens, practitioners, politicians and industry	Stakeholder perspectives on open data usage
Lindman (2014)	Data providers, developers and customers (citizens)	Studying the best business models for open data
Hjalmarsson et al. (2015)	Data providers, third-party developers and open data brokers (mediators)	Understanding the public value of open data for each of the stakeholders
Gonzalez-Zapata and Heeks (2015)	Primary stakeholders as data providers (e.g. politicians, public officials, public sector practitioners, international organizations) and secondary stakeholders as users and contributors (e.g. civil society activists, funding donors, ICT providers, academics)	Understanding bureaucratic, political, technological and economic perspectives of different stakeholders on open data
Susha et al., 2015	Data providers and users	Organizational aspects in engaging users in the open data movement
Ojo et al. (2016)	Data providers and consumers	Openness and usability of open data portals
Charalabidis et al. (2018)	Intermediaries and business players	Generation of open data-driven innovations

Source: own elaboration

closely analyze how they indeed interconnect with each other in a dynamic and networked manner. The biggest gap in the academic background on the topic is that all previous research studies in the area have been concentrated on understanding how different stakeholders view various manifestations of the open data usage; practically with no attention to the question on how all these important actors interplay themselves and collaborate with each other to advance the concept. Taking into account the networking nature of the open data phenomena, this aspect is ripe for further research. This is especially important to examine, since open data is widely regarded as a collaborative movement. In this regard, in contrast to traditional trends in academic literature, the study in this chapter will focus on the analysis of the concept in an unusual context, resorting to multidimensional institutional and stakeholder perspectives in understanding the concept as a key operational platform to promote open data governance and related multi-level channels of political communication which it aims to create. In this respect, the author will try to analyze the promising nuances of political, economic and social relationships that could theoretically emerge in modern digital government ecosystems and related networking platforms between its key players. The generated participatory, civic engagement and peer-to-peer communication channels in open data-driven networks are of special interest in the research since the generalization of hypotheses in this area could provide a fruitful foundation for the further promotion of the concept not only as a technological paradigm but also as an interesting political concept that could *hypothetically* provide new ways to network people in a highly collaborative manner.

3 Understanding the Main Focus of the Research

The collaborative nature of the open data concept is studied in this book firstly through an intrinsically user-centered approach, especially in reviewing all previous literature that theoretically and empirically studied related networking activities among key stakeholders of the open data movement in various parts of the world from various manifestations and perspectives that represent both public and private sectors of the economy. Furthermore, taking into account the multidimensional nature of the phenomena, the analysis of related networking platforms in the area is also presented from various institutional perspectives, especially in regard to the investigation of a wide range of international, regional, national and local initiatives aimed at further diffusion and even globalization of this political concept at a completely new level of collaboration between its main institutional actors. This new institutional trend is actively promoted today by policy-makers, practitioners and non-governmental players at different administrative levels in various parts of the world and is ripe for further research and investigation. In general, the theoretical study presented in this chapter will include three stages of research, using stakeholder and institutional analyses, which themselves seek to synthesize available knowledge in the area and, more importantly, help to postulate new hypotheses and assumptions for actual empirical research and cases studies presented in the following parts of the book.

In this connection, *the stakeholder analysis*, which will be presented in the discussion part right after the case studies on Finland and Sweden, will methodologically be applied to understand, first of all, the perspectives of individual stakeholders in the area such as public agents, citizens, businesses, developers of information and communication technologies, media and non-governmental organizations, especially in an attempt to envisage how they see the new promising opportunities that could be generated in the framework of the open data-driven paradigm and will comprise the first stage of the analysis in the discussion. The second stage of the stakeholder analysis is an attempt to realize the perspectives of various peers, that is, those who are directly engaged in the so-called open data-driven peer-to-peer interactions, firstly, among themselves, which might be helpful in understanding, for example, the promise of civic engagement and e-participation processes that are generated and promoted in the area via such interactions. *The institutional analysis* seeks to understand the third perspective, namely the institutional perspectives of different levels of administrative actors, that is, at national, regional, local and, finally, international levels of open data philosophy that, in turn, could help to better realize a truly multidimensional nature of open data governance through the eyes of all its actors on a wide much more global scale.

4 Understanding Key Research Questions

In general, the multi-institutional and multifaceted nature of interactions between various stakeholders in a wide range of open data systems is tentatively defined in this work as *a multidimensional digital political participation* and conceptually referred by author throughout his research as a key theoretical platform to promote open data governance as a collaborative phenomenon. Therefore, it is quite important to identify, first, the main political implications of such interactions about the perceptions on open data governance by all stakeholders, regardless of cultural and administrative contexts they are implemented in. In this regard, the key areas of the theoretical discussion in both stakeholder and institutional analyses in this research are formulated in the following questions:

1. What is the promise of multi-institutional communication in the realm of open data governance, respectively, from the perspectives of all these three early mentioned actor dimensions (e.g. individual stakeholder, peer-to-peer and administrative institutional ones)?
2. How could such individual, peer-to-peer and institutional perspectives help to network politically various actors around the open data movement?
3. What are the practical implications of the analysis for further empirical research and related case studies presented in the following parts of the work and what hypotheses need to be tested?

5 Understanding the Theoretical Basis of the Research

The research is based on four underlying theories that shape the conceptual framework of the theoretical analysis and empirical case studies on Finland and Sweden to test the discussed hypotheses and assumptions with an ultimate goal to propose a universal analytical framework in studying similar cases in other cross-border and cross-institutional contexts. Given the multi-institutional nature of the open data concept, such theoretical bases as institutional, new public management, stakeholder and structuration theories were used to study this complex phenomenon in all its diversity.

5.1 Understanding the Institutional Theory

The institutional theory is one of the most universally accepted conceptual approaches to analyze a wide range of public structures, rules, norms, symbols and, more importantly, newly established decision-making processes and organizational patterns that emerge within e-government institutions (Kassen, 2019a; Khurshid et al., 2020; Melin, 2016; Zheng et al., 2013), including as part of open data governance paradigm (Janssen et al.,

2012). These institutional and organizational arrangements tend to inform, motivate (Altayar, 2018; Yildiz, 2003), shape and, in most cases, determine the success or failure of related policies and implementation strategies adopted by government officials, practitioners and developers in the area (Ingrams, 2017; Kassen, 2017b; Luna-Reyes & Gil-Garcia, 2011). Due to the fact that decision-makers often have to follow certain fundamental rationale in pursuing these policies and implementation programs (Scott, 2005), it is important to shed light on these deterministic aspects of public sector reforms. The analysis of various components of these processes, regulations and organizations, which turn to be persistent as socioeconomic phenomena, helps to better understand the contexts, directions and even the final goals of related technology-driven public sector reforms. For example, institutional theory postulates that it is important to take into account the political and administrative contexts in analyzing innovative initiatives that indirectly affect the flow of technology-driven reforms in government.

The institutional theory could also be helpful in analyzing hierarchical and networking roles of various actors that drive the open data governance movement as, first of all, interactive and collaborative phenomena, especially in understanding how related aspects of cooperation and collaboration between different institutions could be promoted within various open data-driven platforms and projects initiated by a wide range of actors in the market, both public and private ones (Ruijer et al., 2020; Sandoval-Almazan & Gil-Garcia, 2016). This could be especially important if one seeks to understand the key enablers and barriers that directly affect the adoption of various technology-driven implementation strategies in a multidimensional manner (Zuiderwijk & Janssen, 2014), for example, in understanding complex relationships between information technologies, organizational structures and institutional processes (Hassan & Gil-Garcia, 2008; Karkin et al., 2018). All these institutional contexts, especially if they are analyzed in a cross-border manner, could be useful in generating new knowledge about these decisive environmental components and pressures for reforms (or resistance) in a more comparative and holistic way.

5.2 *Understanding the Theory of New Public Management*

The theory of new public management (NPM) is another conceptual approach that has recently seen revived interest in academic research across various disciplines that study open data governance as a certain playground to experiment in government (Criado et al., 2021; Kassen, 2017c; Kudo, 2016; Longo, 2011; Svärd, 2019) with a wide range of innovative solutions and tactics adopted previously by the private sector management across the world (Przeybilovicz et al., 2018). Indeed, given the ultimate goal of open data philosophy to propagate the values of transparency, collaboration, participation, civic engagement and self-governance in reusing government information for various third-party open data-driven projects, the NPM theory could truly find a new experimental basis to test further its main ideas, especially in regard to the privatization of various e-government services in private and civic sectors of economy. This could become possible due to the democratic nature of open data to emancipate the local business and civic communities by reusing public information for their own purposes, including in developing their own digital services for partners and peers. After all, open data requires a high level of discretion and autonomy in decision-making as well as a certain encouragement of the entrepreneurial approach and competition in the promotion of open data-driven public services.

The concept of open data could change the ways of how public agents and citizens should now communicate with each other, where the latter are viewed not only as clients but more as equal participants of such a complex and sophisticated public corporation as government, especially taking into account that the collaboration between citizens themselves as peers is beginning to play an ever-increasing role in open data-driven projects, which could be developed by technically savvy citizens or independent developers. In this respect, the active participation of civic stakeholders and their engagement in the development of digital public services, the digitalization of which, due to the advance of new government technologies today, greatly simplifies the task of the transition to more effective collaborative models of public service delivery, regardless of whether they are private or public ones, could offer new ways to advance government as a truly transparent institution of decision-making. Various feedback and public relations instruments of communication that are embedded in many open data-driven projects and, most importantly, the cooperation between those who publish datasets and those who reuse them in their own third-party platforms provide a new promising channel to advance the concept further, probing new empirical evidences for its theoretical development and related hypotheses testing.

5.3 *Understanding the Structuration Theory*

Given a truly multidimensional and complex nature of the open data phenomena, the structuration theory could ideally fit the niche for additional theoretical background to support related research on the topic, especially in regard to the advancement of open data through time and space as both a historical manifestation of the socioeconomic development and continuous process of innovation in one given society. The case study research requires the detailed description and integration of various methods of analysis to reveal and investigate all aspects of studied phenomena, more so for such a complex one as open data governance. Taking into account the examination of various contexts and different actors' perspectives incorporated in one research: political, economic, social, technological, administrative and many others, such an inquiry should ideally be done in a cross-disciplinary manner. In this regard, the structuration theory is useful in understanding the ever-changing nature of technology-driven public sector reforms in a strong correlation with the existing social structures and viewpoints of those who are directly engaged in the initiation (Tungela et al., 2018), promotion and, more importantly, adoption (assimilation) of the innovations into established political and socioeconomic structures within studied societies (Guo et al., 2009; Heinze & Hu, 2005; Hossain et al., 2011; Iyamu, 2020; Kassen, 2019b; Puron-Cid, 2013; Rawahi et al., 2016).

The main advantage of the theory is that it helps to study governance empirically in a holistic manner through the analysis of key factors that determine the organization assimilation of the studied complex social phenomenon associated with it (DeSanctis & Poole, 1994; Giddens, 1984; Van Veenstra et al., 2010), resorting to certain semantic flexibility and discourse in analyzing the emergence of complex networking social structures and human relations, the change of which presumably affects its adoption by various target audiences (see Fig. 1.2). By embedding demonstrative cases and perspectives of various stakeholders on them from different institutional and country contexts (Omar et al., 2020), especially in the context of such egalitarian societies as Finland and Sweden, the adoption of the structuration theory to the research could help to understand the complexity of the open data movement as an active living mechanism and, more importantly, offer new ways forward on how to analyze the actions of various agents within it who potentially are able to transform social and government structures through their interactions, observations

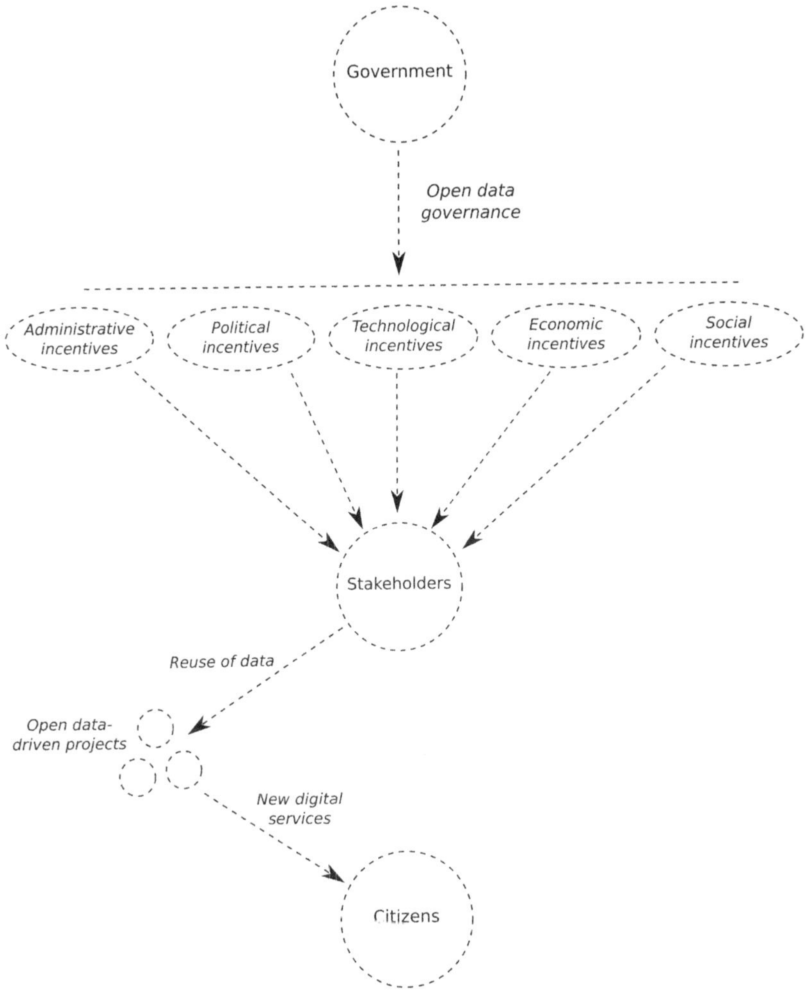

Fig. 1.2 Understanding key aspects of open data governance: looking at the intersection of administrative, political, technological, economic and social dimensions. *Source: own illustration*

and reflections. The combination of various methods of analysis such as semi-structured interviews, document analysis, context and content analysis of key drivers and barriers of the open data movement in each of the studied country cases eventually helps to understand the phenomenon in all its diversity and develop and offer practical recommendations for policymakers and practitioners on how to leverage the related data-driven innovations and promote more effective governance in the area.

5.4 *Stakeholder Theory*

The stakeholder theory is one of the most important conceptual approaches utilized by the research in an attempt to aggregate, analyze and reflect on the perspectives of various actors who are engaged in the advance of open data governance phenomena in each of the country's cases. The viewpoints of such important stakeholders as open data practitioners, businesses, civic communities, independent developers, journalistic and academic communities, non-governmental organizations and thinks tanks, which were collected as a result of field studies in Nordic Europe, allowed to enrich the research with rich empirical data to shed light on the open data governance phenomenon as a collaborative and highly networked movement. Taking into account the ultimate goal of open data to promote greater civic engagement (Burton et al., 2020) and generation of innovations in both public and private sectors of economy thanks to participation of various members of civil society in the process, the stakeholder theory helps to evaluate the adoption of open data governance from various perspectives (Gonzalez-Zapata & Heeks, 2015; Luthfi et al., 2020) to identify and locate key benchmarks, facilitating drivers and obstacles on the road.

The key proponents of the stakeholder theory in the scientific research argue that the interests of those who are associated with or could be affected, in both positive and negative ways, by the studied phenomena should be understood by government policymakers and ideally incorporated into the final equations on strategic decision-making (Flak & Rose, 2005; Kassen, 2021; Luthfi & Janssen, 2021; Scholl, 2001). This prerequisite is also important when studying the perspectives of peers in a cross-institutional analysis of complex government phenomena such as e-government (Pandey & Gupta, 2017). In this regard, the task of the researcher is to identify these interested parties and shed light on their perspectives to see the studied subject in a detailed and comprehensive manner (Freeman, 1999). Such attributes as norms, power and resources, value, legitimacy, urgency, interest, relationship and so on are identified by various scholars as important in locating, identifying and categorizing key stakeholders for the analysis (Balta et al., 2015; Mitchell et al., 1997; Rose et al., 2018; ul Hassan & Curry, 2021).

Open data is a complex multidimensional collaborative process, the key elements of which are deeply embedded in the fabric of society. In this regard, relevant research should be accompanied by flexible and hybrid methodological approaches that can help to understand the phenomena from a variety of institutional points of view. Thus, the combination of these four theoretical approaches provides a multidimensional conceptual framework for a more holistic and systematic analysis of open data phenomena. For example, when adopting institutional theory, it is interesting to understand the implications of surrounding political, economic, social, technological and international contexts, while structuration theory could be extremely useful for generalizing and synthesizing knowledge derived from policy analysis in a more holistic manner, taking into account the layered structure of open data movement. Finally, new public management and stakeholder theories could be useful in tracking the emergence of new promising socioeconomic and behavioral trends in communication between public, non-governmental and private stakeholders around the phenomenon of open data, especially when analyzing how various actors such as citizens, businesses, independent developers, charities, academic and journalistic communities from non-government and private sectors could interact with each other and equally contribute to the development and promotion of various platforms based on open data. In general, the combination of these four conceptual frameworks can be universally used in a repetitive manner in other related case studies, regardless of cultural or national context.

6 Understanding the Country Case Selection: Focus on Finland and Sweden

The main reason to choose these two nations for the case study analysis of open data governance is unique political and socioeconomic contexts that these countries could provide for the investigation. The adherence to main principles of social and parliamentary democracy (Alestalo & Kuhnle, 1986; Arter, 1999) allowed to create one of the most egalitarian societies in the world in Nordic Europe. Transparency of public information is a crucial public value that helps to build open societies. In this regard, both Finland and Sweden are famous for the popularity of open knowledge and open science movements. Moreover, it is necessary to note that these two Nordic nations are a homeland for the first freedom of information law in the history of humankind (Blanton, 2002). The two-century-old traditions of freedom of information legislation and social democracy in the political arena and modern focus on the development of telecommunication industries in economy are what make Finland and Sweden so similar in their context. The popularity of open knowledge, open source software movements and even pirate politics, which provide extremely favorable conditions to promote open data-driven philosophy of transparency and electronic participation, is one more reason to regard these country cases appealing for the analysis in an attempt to understand the effects of various political and socioeconomic contexts on the development of open data governance philosophy, which is currently being actively promoted in Nordic Europe. Therefore, these two nations present a really interesting venue for the analysis of these new phenomena from the unique political, socioeconomic and technological perspectives (Fig. 1.3).

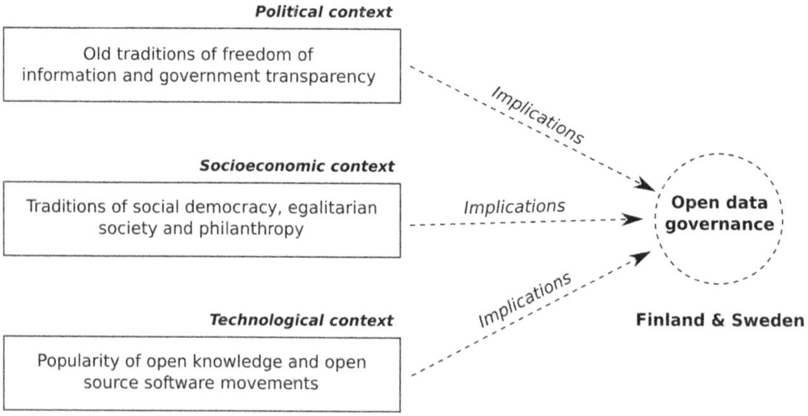

Fig. 1.3 Understanding the implications of political, socioeconomic and technological contexts in Finland and Sweden. *Source: own illustration*

7 Understanding Theoretical Part of the Research: Stakeholder and Institutional Analysis

The first and foremost important and crucial step in any case study is the introduction to the main idea of the research question (Eisenhardt, 1989; Yin, 2013). In this regard, the concept of open data governance is explained theoretically in Chaps. 2 and 3 of the research with a detailed analysis of all previous related studies in the area in regard to various stakeholder and institutional perspectives on open data governance as a collaborative phenomenon. The topic is reviewed in terms of public values that the reuse of public information and raw datasets presumably brings to both government and private sectors of national and local economy such as the growth of public sector innovations (Janssen et al., 2012; Kassen, 2016; Lakomaa & Kallberg, 2013; Yavuz et al., 2020), the increased level of political communication (İkiz et al., 2014), democratic participation and civic engagement (Davies, 2010; De Cindio, 2012; Lathrop & Ruma, 2010), the collaborative activity of public-private partnerships in the area among key stakeholders (Linders, 2012; Ojo & Janssen, 2013), the creation of independent networks of developers and technically savvy citizens (Hartung et al., 2010; Kassen, 2020; Kuk & Davies, 2011) and so on. In this regard, the conceptual and operational definitions of open data and open data governance concepts will be provided throughout these chapters in the theoretical analysis.

One of the most important steps in preparing for the case study research, especially in investigating political dimensions of the open data phenomena, was the institutional and stakeholder analysis. In this regard, it was important to provide a general overview of previous academic and professional research on the topic that could help to compare them later with key findings in the second major more empirical part of the investigation, which was based on the analysis of rich empirical data received from documents analysis of official texts in various political strategies and legal acts, speeches of public officials and politicians in the area and, more importantly, semi-structured interviews, focus group discussions, content analysis and observation of actual open data-driven projects and startups created by citizens themselves in the area. It is especially true when it comes to the discussions of generalizations on typical driving forces and barriers as well as on the role of key stakeholders associated with the adoption of the open data governance concept, which were themselves derived from real practices, especially in promoting open data-driven public sector reforms and related implementation strategies and policies.

In this respect, the primary goal of the institutional and stakeholder analysis was to update available theories and assumptions by resorting to multiple case studies in a consistent and successive manner and cross national comparative analysis of its main findings. The cases of Finland and Sweden provided interesting narratives of open data governance in terms of the realization of official projects and platforms where the role of national and local governments varies. Likewise, the participation of representatives from the non-governmental sector, local businesses, developers and other important stakeholders changes greatly across the nations, too. Sharing some similarities in country backgrounds, especially in terms of political and socioeconomic structures, as well as demonstrating a decisive role of central government in promoting various technology-driven public sector reforms and overall political importance of the e-government topic in the public agenda, these nations provide an interesting playground to test previously proposed assumptions about typical driving forces, challenges and dynamics of the open data movement and understand the implications of political, socioeconomic and technological contexts on its development.

8 Understanding the Empirical Part of the Research: Case Study on Finland and Sweden

The case study itself, which was consistently replicated in two country contexts in a successive order in relationship to Finland and Sweden, consisted of seven key methodological steps such as the lead-in part or introduction, which is dedicated to the brief overview of the open data movement in one given country, the short literature review on the development of open data governance in each of these two nations, the context analysis of the country background, that is, the study of unique political and socioeconomic environments that are presumably conducive for the development of the open data concept in these societies, the analysis of the freedom of information laws related to open data adopted by national authorities, the network analysis of how the open data concept is diffusing among its key stakeholders and technology entrepreneurs, the policy analysis of key trends in decisions-making in the area and analysis of how open data governance is promoted in various sectors of economy and finally the content analysis of official open data platforms and independent open data-driven projects. The system and comparative analysis of generalizations and key findings on the open data movements in each of the countries were provided in the conclusion. The material for the empirical part of the research was mostly collected as the results of multiple field studies which were organized by the author in Nordic Europe in 2015–2017 (e.g. semi-structured interviews and focus group studies), document and content analysis of sources from various government bodies, international organizations and statistical agencies in the area. In this regard, the data for content analysis was investigated in the research through the review of related regulatory acts, official statistics and social media channels and analysis of various open government portals and independent open data-driven projects.

I. Key regulation documents:

In Finland:

1. The Act on the Publicity of Official Documents (1951).
2. The Act on the Openness of Government Activities (1999).
3. Open Data Program of Finland. Ministry of Finances (2014).

In Sweden:

1. The Act on the Freedom of Publishing and the Right of Access to Official Documents (1766).
2. Sweden's Open Government Partnership Action Plan 2014–2016 (2012).
3. The Swedish Personal Information Act (1973).

II. Key sources of official statistics:

(A) In Finland:

Government sources:

1. The Ministry of Transport and Communications of Finland. http://www.lvm.fi/home
2. The Ministry of Justice of Finland. http://oikeusministerio.fi/en/
3. The Ministry of Finance of Finland. http://vm.fi/en/frontpage
4. The Statistical Agency of Finland. http://www.stat.fi/index_en.html

Independent sources:

1. The National Land Survey of Finland. http://www.maanmittauslaitos.fi/en/contacts
2. The Finnish Meteorological Institute. http://en.ilmatieteenlaitos.fi/
3. The Forum Virium Platform. https://www.forumvirium.fi/en
4. The Open Knowledge Finland. http://fi.okfn.org/about/

(B) In Sweden:
Government sources:

1. The Swedish innovation agency—Vinnova. http://www.vinnova.se/en/
2. The Swedish Innovation Center. http://swedsoft.se/
3. The Swedish Ministry of Enterprise and Innovations.

Independent sources:

1. The Kista Science City. www.kista.com
2. The Swedish Institute of Computer Science. https://www.sics.se/about-sics
3. Linköping University. http://www.liu.se/?l=en
4. The Research Institute Acreo. https://www.acreo.se
5. Lund University Internet Institute. http://luii.lu.se/research/legal-solutions-for-open-data

III. Key empirical cases:
(A) In Finland:
The key national open data platforms:

1. Open Ministry. http://openministry.info
2. Avoin Ministerio. www.avoinministerio.fi
3. Avoindata. The national open data project of Finland. https://www.avoindata.fi/en

The key local open data platforms:

1. The Helsinki Region Infoshare. http://www.hri.fi/en/

2. Tampere Avoin Data. http://www.tampere.fi/tampereen-kaupunki/tietoa-tampereesta/avoin-data.html
3. Oulu Avoin Data. http://www.ouka.fi/avoindata
4. Jyväskylä Avoin Data. http://data.jyvaskyla.fi/

The key networking platforms:

1. The Civic Tech Sweden Workshops. https://civictech.se/en/events
2. The 6Aika project. http://6aika.fi/in-english
3. The Sitra Finnish Innovation Fund. http://www.sitra.fi/en/about-sitra
4. The Business innovations networking platform—Protomo. www.protomo.fi
5. The peer-to-peer innovations generating community platform—Ominvoimin. http://ominvoimin.com
6. The Valtori Government ICT Center. http://www.valtori.fi/en-US
7. The Helsinki Loves Developers Project. http://dev.hel.fi/
8. The Comprehensive Knowledge Archive Network. http://ckan.org/about/
9. The Open Finland Challenge. http://openfinlandchallenge.fi/en/
10. The Ultrahack networking platform. https://ultrahack.org/
11. The Apps4Finland networking platform. http://www.apps4finland.fi/

The key independent open data-driven platforms:

1. The rOpenGov initiative. The data-driven collaborating and networking platforms for software developers. http://ropengov.org/
2. The Detademo project. The participatory budgeting platform. http://datademo.fi/english/
3. The Cloud'N'Sci.fi project. https://cloudnsci.fi/
4. The Agience project. http://agience.fi/
5. The Puoluekentta project. http://puoluekentta.tstm.info/
6. The Parliamentary Explorer project. https://www.microsoft.com/en-us/store/apps/eduskunta-explorer/9nblggh085ww
7. The Päätökset project. The government decision-making tracking project. http://dev.hel.fi/paatokset/

(B) In Sweden:
The key national open data platforms:

1. The official open data project of Sweden. http://oppnadata.se/
2. The Open Government Tracking Project. http://www.opengov.se

The key local open data platforms:

1. The official open data project of Stockholm. http://dataportalen.stockholm.se/dataportalen/
2. The official open data project of the city of Lund. http://www.lund.se/psidata
3. The official open data project of the city of Uppsala. https://www.uppsala.se/psidata

The key networking platforms:

1. The TechMeetups Stockholm. http://www.meetup.com/Stockholm-TechMeetups/
2. The Swedish Startup Space. http://swedishstartupspace.com
3. The Popdevelop networking project. Malmo, Sweden. http://popdevelop.com/
4. The Meta Solutions Project. http://metasolutions.se/tag/project/
5. The Kids Hack Day. http://www.kidshackday.com/
6. The Tele2 Hack Day. https://tele2.confetti.events/
7. The Health Hack Day. http://healthhackday.com/
8. The Hackaway project. http://www.hackaway.com/
9. The Nordic Internet of Things Hackathon. http://nordiciothackathon.com/
10. The Hack for Sweden. http://hackforsweden.se/
11. The East Sweden Hack. http://eastswedenhack.se/west-coast-big-apps
12. The Open Stockholm Award. http://www.openstockholmaward.se

The key independent open data-driven platforms:

1. The Trafiklab project. The open data-driven e-transportation project. https://www.trafiklab.se

2. The Handlingar project. The FOIA requests platform. https://handlingar.se/
3. The Journalism++ Project. http://jplusplus.se/
4. The OpenAid project. http://openaid.se
5. The Lagen open legislation project. www.lagen.nu
6. The Swedish National Data Service. https://snd.gu.se/en

9 Understanding the Main Directions of the Research

Taking into account that the main principle of open data is its reuse and further processing by the members of civil society, this research is aimed at furthering research about this promising direction of the concept. Open data governance is beginning to play an ever-increasing role in public sector reforms all over the world. Likewise, the research on open government as a theoretical concept is on the rise in many fields. However, related literature rarely focuses on the analysis of empirical evidence from real practices and even less from both political and socioeconomic perspectives. In this connection, the main purpose of the book is to understand the dynamics of open data governance in the contexts of high developed egalitarian societies and reveal an important role of political and socioeconomic contexts for its key stakeholders by resorting to an interesting case study of open data movements in Finland and Sweden, which have recently adopted various networking initiatives in the area (see Fig. 1.4).

In this regard, the theoretical (e.g. introduction to the concept of open data governance through stakeholder and institutional analysis) and empirical parts of the research (which were analyzed from the perspectives of two country contexts, e.g. through the eyes of various actors, institutions, a wide range of public, private and civic open data platforms and multiple case studies of data-driven collaboration, participation and networking in Finland and Sweden) were reflected in the corresponding structure of the book with the analysis of key drivers and challenges of the open data movement and were further discussed and elaborated in the conclusion chapter.

I. Theoretical part of the research:

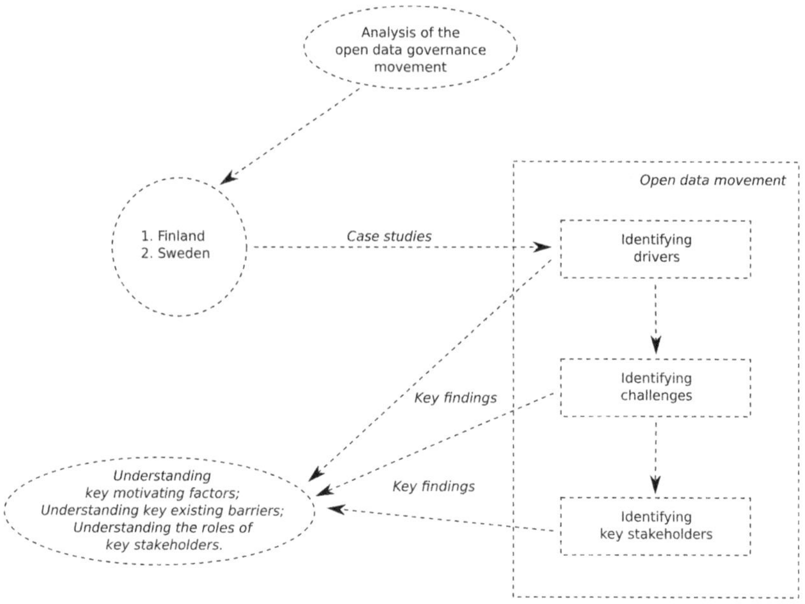

Fig. 1.4 The general outline of key directions of the research. *Source: own illustration*

- Chapter 1 (introduction to the conceptual and methodological background);
- Chapter 2 (stakeholder analysis of the open data governance);
- Chapter 3 (institutional analysis of open data governance);

II. Practical part of the research:

- Chapter 4 (empirical analysis of open data governance in Finland);
- Chapter 5 (empirical analysis of open data governance in Sweden);
- Chapter 6 (further discussion of key findings and conclusion).

The successful development of various open data-driven projects and related civic startups in the area and the overall active development of the concept in these nations at both central and local levels of governance as well as the recent emergence of really promising networks of political communication between its main stakeholders are ripe for research and analysis from multitude of perspectives and as such should be regarded as new interesting benchmarks of political maturity of civil society. In this regard, the results of the investigation that are based on case studies and comparative analysis of open data policies adopted in each of the countries could provide a better conceptual framework and platform for future studies in the area from an intrinsically political perspective. In this regard, it would be interesting to analyze how the open data concept is promoted and realized in unique cultural and political contexts in Nordic Europe in order to update existing theoretical foundations in the area and generate new knowledge, for example, through comparing the results of the research with previous studies and generalizations on existing open data practices in other regions of the world. In this regard, in analyzing key drivers and challenges of the concept adoption, it is important to understand the direct and indirect implications of cultural, political, economic and other contexts on the diffusion of the phenomena so that to learn from the best practices in the area and comprehend better the multidimensional and collaborative nature of open data governance, which could change the fundamentals of public sector reforms and increase transparency of government institutions, ideally in other countries around the world, too. In general, the multidimensional comparative analysis could help to study the similarities and differences of the concept adoption from regional or even global perspectives, in an attempt to contribute to the general knowledge about open data as a presumably context-dependent phenomenon.

References

Alestalo, M., & Kuhnle, S. (1986). 1 the Scandinavian route: Economic, social, and political developments in Denmark, Finland, Norway, and Sweden. *International Journal of Sociology, 16*(3–4), 1–38.

Altayar, M. S. (2018). Motivations for open data adoption: An institutional theory perspective. *Government Information Quarterly, 35*(4), 633–643.

Arter, D. (1999). Party system change in Scandinavia since 1970: 'Restricted change' or 'general change'? *West European Politics, 22*(3), 139–158.

Balta, D., Greger, V., Wolf, P., & Krcmar, H. (2015, January). E-government stakeholder analysis and management based on stakeholder interactions and resource dependencies. In 2015 48th Hawaii International Conference on System Sciences (pp. 2456–2465). IEEE.

Blanton, T. (2002). The world's right to know. *Foreign Policy, 1*, 50–58.

Burton, P., Tiernan, A., Wolski, M., Drennan, L., & Morrissey, L. (2020). *Resilient cities, user-driven planning, and open data policy. In open cities| open data* (pp. 383–400). Palgrave Macmillan.

Charalabidis, Y., Zuiderwijk, A., Alexopoulos, C., Janssen, M., Lampoltshammer, T., & Ferro, E. (2018). *Open data evaluation models: Theory and practice. In the world of open data* (pp. 137–172). Springer.

Criado, J. I., Dias, T. F., Sano, H., Rojas-Martín, F., Silvan, A., & Filho, A. I. (2021). Public innovation and living labs in action: A comparative analysis in post-new public management contexts. *International Journal of Public Administration, 44*(6), 451–464.

Davies, T. (2010). Open data, democracy and public sector reform. A look at open government data use from data. gov. uk, 1–47.

De Cindio, F. (2012). Guidelines for designing deliberative digital habitats: Learning from e-participation for open data initiatives. *The Journal of Community Informatics, 8*(2).

DeSanctis, G., & Poole, M. S. (1994). Capturing the complexity in advanced technology use: Adaptive structuration theory. *Organization Science, 5*(2), 121–147.

Eisenhardt, K. M. (1989). Building theories from case study research. *Academy of Management Review, 14*(4), 532–550.

Flak, L. S., & Rose, J. (2005). Stakeholder governance: Adapting stakeholder theory to e-government. *Communications of the Association for Information Systems, 16*(1), 31.

Freeman, R. E. (1999). Divergent stakeholder theory. *Academy of Management Review, 24*(2), 233–236.

Giddens, A. (1984). *The constitution of society: Outline of the theory of structuration*. University of California Press.

Gonzalez-Zapata, F., & Heeks, R. (2015). The multiple meanings of open government data: Understanding different stakeholders and their perspectives. *Government Information Quarterly, 32*(4), 441–452.

Guo, X., Zhang, N., & Chen, G. (2009). Adoption and penetration of e-government systems: conceptual model and case analysis based on structuration theory. DIGIT 2009 Proceedings, 1.

Hartung, C., Lerer, A., Anokwa, Y., Tseng, C., Brunette, W., & Borriello, G. (2010, December). Open data kit: tools to build information services for developing regions. In Proceedings of the 4th ACM/IEEE international conference on information and communication technologies and development (pp. 1–12).

Hassan, S., & Gil-Garcia, J. R. (2008). Institutional theory and e-government research. In Handbook of research on public information technology (pp. 349–360). IGI Global.

Heinze, N., & Hu, Q. (2005). e-Government research: a review via the lens of structuration theory. PACIS 2005 Proceedings, 75.

Hjalmarsson, A., Johansson, N., & Rudmark, D. (2015, January). Mind the gap: exploring stakeholders' value with open data assessment. In 2015 48th Hawaii international conference on system sciences (pp. 1314–1323). IEEE.

Hossain, M. D., Moon, J., Kim, J. K., & Choe, Y. C. (2011). Impacts of organizational assimilation of e-government systems on business value creation: A structuration theory approach. *Electronic Commerce Research and Applications, 10*(5), 576–594.

İkiz, O. O., Sobaci, M. Z., Yavuz, N., & Karkin, N. (2014, October). Political use of Twitter: The case of metropolitan mayor candidates in 2014 local elections in Turkey. In Proceedings of the 8th international conference on theory and practice of electronic governance (pp. 41–50).

Ingrams, A. (2017). The transparency performance puzzle: A fuzzy set qualitative comparative analysis of policy failure in open government initiatives. *Information Polity, 22*(1), 25–39.

Iyamu, T. (2020). Examining E-government enabling of E-health service through the lens of structuration theory. *International Journal of Sociotechnology and Knowledge Development (IJSKD), 12*(3), 26–40.

Janssen, M., Charalabidis, Y., & Zuiderwijk, A. (2012). Benefits, adoption barriers and myths of open data and open government. *Information Systems Management, 29*(4), 258–268.

Karkin, N., Yavuz, N., Cubuk, E. B. S., & Golukcetin, E. (2018, May). The impact of ICTs-related innovation on public values in public sector: a meta-analysis. In Proceedings of the 19th Annual International Conference on Digital Government Research: Governance in the Data Age (pp. 1–9).

Kassen, M. (2016). A content analysis of key e-government projects in Kazakhstan. In *E-government in Kazakhstan* (pp. 54–65). Routledge.

Kassen, M. (2017a). Open data in Kazakhstan: Incentives, implementation and challenges. *Information Technology & People, 30*(2), 301–323.

Kassen, M. (2017b). E-government politics as a networking phenomenon: Applying a multidimensional approach. *International Journal of Electronic Government Research (IJEGR), 13*(2), 18–46.

Kassen, M. (2017c). Understanding transparency of government from a Nordic perspective: Open government and open data movement as a multidimensional collaborative phenomenon in Sweden. *Journal of Global Information Technology Management, 20*(4), 236–275.

Kassen, M. (2018). Adopting and managing open data: Stakeholder perspectives, challenges and policy recommendations. *Aslib Journal of Information Management, 70*(5), 518–537.

Kassen, M. (2019a). Open data politics: Building a research framework. In *Open data politics* (pp. 1–18). Springer.

Kassen, M. (2019b). *Open data politics: A case study on Estonia and Kazakhstan*. Springer.

Kassen, M. (2020). E-participation actors: Understanding roles, connections, partnerships. *Knowledge Management Research and Practice, 18*(1), 16–37.

Kassen, M. (2021). Understanding motivations of citizens to reuse open data: Open government data as a philanthropic movement. *Innovations, 23*(1), 44–70.

Khurshid, M. M., Zakaria, N. H., Rashid, A., Ahmad, M. N., Arfeen, M. I., & Faisal Shehzad, H. M. (2020, September). Modeling of open government data for public sector organizations using the potential theories and determinants—a systematic review. In Informatics (Vol. 7, No. 3, p. 24). Multidisciplinary Digital Publishing Institute.

Kitchin, R. (2014). *The data revolution: Big data, open data, data infrastructures and their consequences.* Sage.

Kudo, H. (2016). Still convinced by New Public Management or introducing somehow New Public Governance? How Japanese local governments are coping with these changes to deliver public services. Zeitschrift für öffentliche und gemeinwirtschaftliche Unternehmen: ZögU/Journal for Public and Nonprofit Services, 77–88.

Kuk, G., & Davies, T. (2011). The roles of agency and artifacts in assembling open data complementarities. International Conference on Information Systems 2011. Shanghai, China.

Lakomaa, E., & Kallberg, J. (2013). Open data as a foundation for innovation: The enabling effect of free public sector information for entrepreneurs. *IEEE Access, 1*, 558–563.

Lathrop, D., & Ruma, L. (2010). *Open government: Collaboration, transparency, and participation in practice.* O'Reilly Media, Inc.

Linders, D. (2012, June). How can open development improve the effectiveness of aid? Leveraging open data, open standards, and web 2.0 interactivity for better development outcomes. In Proceedings of the 13th Annual International Conference on Digital Government Research (pp. 155–164).

Lindman, J. (2014). Similarities of open data and open source: Impacts on business. *Journal of Theoretical and Applied Electronic Commerce Research, 9*(3), 46–70.

Longo, J. (2011). # OpenData: Digital-era governance thoroughbred or new public management Trojan horse? *Public Policy & Governance Review, 2*(2), 38.

Luna-Reyes, L. F., & Gil-Garcia, J. R. (2011). Using institutional theory and dynamic simulation to understand complex e-government phenomena. *Government Information Quarterly, 28*(3), 329–345.

Luthfi, A., & Janssen, M. (2021). A stakeholders taxonomy for opening government data decision-making. In *International symposium on business Modeling and software design* (pp. 384–391). Springer.

Luthfi, A., Janssen, M., & Crompvoets, J. (2020). Stakeholder tensions in decision-making for opening government data. In *International symposium on business Modeling and software design* (pp. 331–340). Springer.

Martin, M., Kaltenböck, M., Nagy, H., & Auer, S. (2011, June). The Open Government Data Stakeholder Survey. In OKCon.

Melin, U. (2016, August). Challenges and Benefits in an Open Data Initiative–Local Government Case Study of Myths and Realities. In 15th IFIP Electronic Government and the 8th Electronic Participation Conference (EGOV ePart 2016) (Vol. 23, pp. 111–122).

Mitchell, R. K., Agle, B. R., & Wood, D. J. (1997). Toward a theory of stakeholder identification and salience: Defining the principle of who and what really counts. *Academy of Management Review, 22*(4), 853–886.

Ojo, A., & Janssen, M. (2013, June). Aligning core stakeholders' perspectives and issues in the open government data community. In Proceedings of the 14th Annual International Conference on Digital Government Research (pp. 293–294).

Ojo, A., Porwol, L., Waqar, M., Stasiewicz, A., Osagie, E., Hogan, M., ... & Zeleti, F. A. (2016, October). Realizing the innovation potentials from open data: Stakeholders' perspectives on the desired affordances of open data environment. In Working Conference on Virtual Enterprises (pp. 48–59). Springer, Cham.

Omar, A., El-Haddadeh, R., & Weerakkody, V. (2020). Exploring digitally enabled service transformation in the public sector: Would institutional and structuration theory concepts keep the research talking? In *Open government: Concepts, methodologies, tools, and applications* (pp. 911–926). IGI Global.

Pandey, V., & Gupta, S. (2017). Understanding G2G e-government project impasse: A stakeholder theory perspective. *Information Development, 33*(4), 361–374.

Przeybilovicz, E., Cunha, M. A., Macaya, J. F. M., & Albuquerque, J. P. D. (2018). A Tale of two" Smart Cities": Investigating the echoes of new public management and governance discourses in Smart City Projects in Brazil.

Puron-Cid, G. (2013). Interdisciplinary application of structuration theory for e-government: A case study of an IT-enabled budget reform. *Government Information Quarterly, 30*, 46–58.

Rawahi, K. A., Coombs, C., & Doherty, N. (2016). The realization of public value through e-government: a structuration perspective. Proceedings of the 37th international conference on information systems, Dublin, December 11–14 , 2016.

Rose, J., Flak, L. S., & Sæbø, Ø. (2018). Stakeholder theory for the E-government context: Framing a value-oriented normative core. *Government Information Quarterly, 35*(3), 362–374.

Ruijer, E., Détienne, F., Baker, M., Groff, J., & Meijer, A. J. (2020). The politics of open government data: Understanding organizational responses to pressure for more transparency. *The American Review of Public Administration, 50*(3), 260–274.

Safarov, I., Meijer, A., & Grimmelikhuijsen, S. (2017). Utilization of open government data: A systematic literature review of types, conditions, effects and users. *Information Polity, 22*(1), 1–24.

Sandoval-Almazan, R., & Gil-Garcia, J. R. (2016). Toward an integrative assessment of open government: Proposing conceptual lenses and practical components. Journal of Organizational Computing and Electronic Commerce, 26(1–2), 170–192.

Scholl, H. J. (2001). Applying stakeholder theory to e-government. In *Towards the E-society* (pp. 735–747). Springer.

Scott, W. R. (2005). Institutional theory: Contributing to a theoretical research program. *Great Minds in Management: The Process of Theory Development, 37*(2), 460–484.

Susha, I., Grönlund, Å., & Janssen, M. (2015). Organizational measures to stimulate user engagement with open data. *Transforming Government: People, Process and Policy, 9*(2), 181–206.

Svärd, P. (2019). The impact of new public management through outsourcing on the management of government information: The case of Sweden. *Records Management Journal, 29*(1–2), 134–151.

Tungela, N., Mutudi, M., & Iyamu, T. (2018, October). The Roles of E-Government in Healthcare from the Perspective of Structuration Theory. In 2018 Open Innovations Conference (OI) (pp. 332–338). IEEE.

Ul Hassan, U., & Curry, E. (2021). Stakeholder analysis of data ecosystems. *The Elements of Big Data Value, 21*, 1.

Van Veenstra, A. F., Janssen, M., & Tan, Y. H. (2010, August). Towards an understanding of E-government induced change–drawing on organization and structuration theories. In International Conference on Electronic Government (pp. 1–12). Springer, Berlin, Heidelberg.

Yavuz, N., Karkın, N., & Çubuk, E. B. S. (2020). *Explaining government crowdsourcing decisions: A theoretical model. In digital government and achieving E-public participation: Emerging research and opportunities* (pp. 159–183). IGI Global.

Yildiz, M. (2003, May). Examining the motivations for e-government from an institutional theory perspective: evidence from turkey. In Proceedings of the 2003 annual national conference on Digital government research (pp. 1–4).

Yin, R. K. (2013). Validity and generalization in future case study evaluations. *Evaluation, 19*(3), 321–332.

Zheng, D., Chen, J., Huang, L., & Zhang, C. (2013). E-government adoption in public administration organizations: Integrating institutional theory perspective and resource-based view. *European Journal of Information Systems, 22*(2), 221–234.

Zuiderwijk, A., & Janssen, M. (2014, June). The negative effects of open government data-investigating the dark side of open data. In Proceedings of the 15th Annual International Conference on Digital Government Research (pp. 147–152).

CHAPTER 2

Open Data from the Perspectives of Individual Actors of Political Communication

Abstract In analyzing theoretically open data as, first of all, part of an emerging political concept, especially from the perspectives of its main actors, namely policymakers, practitioners and technology-savvy enthusiasts, that is, those who are usually expected to promote the public agendas of open data governance, it is necessary to mention, firstly, the three classic domains of related political communication channels, which are presumably generated in related digital government transactions such as public or government, civic and business-centric perspectives, and then three non-standard dimensions such as those that could exist from the perspectives of, respectively, independent developers, media communities and non-governmental institutions (e.g. public activists, think tanks, academia and philanthropic organizations).

Keywords Open data • Stakeholders • Public sector • Citizens • Businesses

M. Kassen, *Open Data Governance and Its Actors*, Studies in National Governance and Emerging Technologies,
https://doi.org/10.1007/978-3-030-92065-4_2

1 The Public Sector Perspective: Is It Still All About Government-Centric Approach?

The public- or government-centric perspective has been probably one of the most popular approaches in studying open data since the emergence of the concept in the political agenda in early 2010s (Conradie & Choenni, 2014; Kassen, 2013; Ojo & Janssen, 2013), given that it claims to create, first of all, a cost-effective public administration ecosystem (Pereira et al., 2017). The public perspective is widely regarded as a classical one not only in academic circles (Kitchin, 2014) but, more importantly, in the professional community of public administrators and e-government practitioners around the world as well as among economists in the e-commerce sector, where, in fact, the related pre-e-government research and its nomenclature paradigm appeared first (Helbig et al., 2009), particularly in FinTech industries (Kassen, 2019a; Tinholt, 2013; Yavuz & Saiti, 2020) in a terminological attempt to distinguish the public and private actors in government-related networking and communication (Boughzala et al., 2015; Fusi & Feeney, 2020; Kassen, 2016) and only then it was extrapolated, first, into e-government and later related open data literature, especially, in such popular areas of research as, for example, finding the linkage between trust and confidence in government (Gonzálvez-Gallego & Nieto-Torrejón, 2021; Meijer et al., 2014; O'Hara, 2012), or measuring the practical aspects of net-usability among government users in various public electronic systems, the related definition framework for which had been generated earlier in the corporate sector. In its original sense, the government perspective of data-driven public administration interactions includes such key players as government itself, business and citizens, while the relationships themselves are usually indicated in a strictly dyadic manner.

In regard to the open data movement that could presumably be generated only in related modern digital projects and startups; this classical government-centric approach could be adapted in an attempt to understand political relationships between its key players, too. Especially, taking into account that government is usually a key initiator and promoter of all related civic and business engagement policies in the area, letting alone the fact that it is itself a solely publisher of all datasets, which are necessary for the normal functioning and systematic updating of various third-party open data-driven projects, startups and applications. Therefore, in order to better realize the government-centric perspective, one should imagine graphically a very straightforward organizational process of open data-driven collaboration where the public sector performs the first and foremost important step by just publishing datasets. Consequently, citizens and businesses should take the second step by reusing them in their own open data-driven products. In this regard, the process could be regarded successful only if it results in generating data-driven public sector innovations, that is, the third final stage should bring public benefits or public values such as, for example, new e-government or e-commerce open data-driven services, initiated respectively by citizens and businesses in areas which are interesting for them. Needless to say that the reuse of open data might not always result in the direct manifestations since datasets could also be used for personal academic, journalistic or corporate purposes (see Fig. 2.1).

Moreover, taking into account the networking nature of open data technologies, this purely public paradigm could also now incorporate such non-traditional subjects as, for example, non-governmental organizations, developers and even mass media, whose participation in the interactions became especially apparent in the light of recent diffusion of such arguably

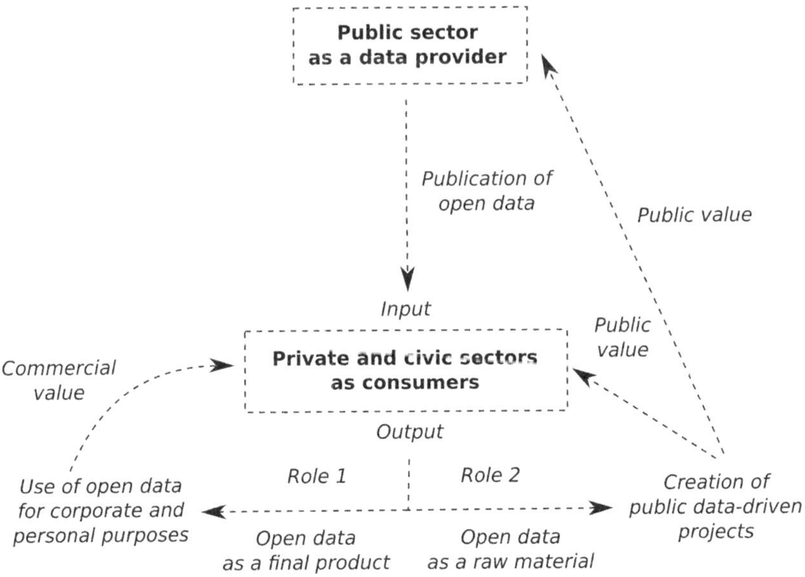

Fig. 2.1 The provision of public data as the first stage in the promotion of open data governance. (Source: Own illustration)

independent collaborative platforms as open data-driven civic engagement and e-participation projects, for example, by using various blockchain technologies (Joseph, 2019; Kassen, 2021a). In this regard, the government-centric perspective could also *hypothetically* generate such new promising dyadic political networks as *government-to-NGOs*, indicating the ever-increasing role of the non-government sector in the promotion of the open data phenomena as a universal public value; *government-to-developers*, reflecting the recent trend in engaging independent computer programmers and experts and even, sometimes, technically savvy citizens to the collaborative development, testing and launching of various open data-driven projects; and *government-to-media*, emphasizing an important role of mass media as a bridge in public relations between government, citizens, non-governmental players, developers and businesses in the area.

In this regard, taking into account this public sector-centered perspective, open data governance as a theoretical concept could be described as a political platform aimed at providing government information (e.g. datasets, maps, archives, official statistics, geostationary data, video and audio files, paper and electronic records) to citizens, developers, businesses, media and even other government agencies with an ultimate goal to ensure the transparency of government itself and, more importantly, *promote the reuse of open data in third party e-government and e-commerce solutions and applications.* Even if it is sometimes just a two-way process of communication, the main role of government, in this respect, should be regarded as crucial as never before. In this connection, the primary focus of open data-driven activities in this direction is the reformation of the public administration system itself; therefore, the public sector should be theoretically considered as a focal point of all transactions and a key player, which incorporates all open data projects and networks and integrates all other subjects of related political communication processes as a certain *supernode* and not just a passive data provider

Research question # 1:
By initiating various open data government projects and providing various types of datasets in a transparent manner via special freely accessible repositories, the public sector presumably sends a strong political signal for other stakeholders to join the movement and motivates them to reuse open data. In this regard, it seems interesting to understand how the government acts as a supernode in uniting the interests of various actors and creating new promising political networks between them.

2 The Citizen's Perspective: Is Civic Engagement an Ultimate Recipe to Build Independent Open Data Governance?

Conceptually, the civic perspective could probably be one of the most promising viewpoints in understanding the idea of open data governance as a presumably bottom-up collaborative phenomenon since it emphasizes the central role of citizens as a crucial component of all open data-related relations and activities, especially, taking into account that it could offer arguably mutually beneficial dividends for both citizenry and public sector (Ciociola & Reggi, 2015; Janssen et al., 2012; Kassen, 2019b; Kassen, 2020; Puussaar et al., 2018; Ricker et al., 2020; Sieber & Johnson, 2015; Weerakkody, Irani, et al., 2017; Weerakkody, Kapoor, et al., 2017), dramatically expanding the public administration-focused approach of the classical government-centered perspective, which is still quite popular among many practitioners and academics in the field, due to a simple fact that government, in many cases, itself is an initiator and promoter of almost all technology-driven public sector reforms. However, recognizing an obvious key role of government agencies in these communication processes, many e-government scientists begin to elaborate the idea further in accentuating on an ever-increasing role of civil society, which helps to develop new dimensions of open data governance paradigms, such as government 3.0 (Ojo & Millard, 2017; Yli-Huumo et al., 2018) or "collaborative governance" (Kassen, 2021b; Meijer & Potjer, 2018; Susha & Gil-Garcia, 2019). In this regard, in accordance with the established dyadic system of description, all open data-driven political interactions and communication networks, in this respect, could presumably be subdivided into the following dyadic pairs: citizens-to-government (e.g. in seeking for initial consolatory support), citizens to businesses (e.g. in seeking for initial funding and partnership), citizens-to-NGOs (e.g. in seeking for much-needed organizational assistance), citizens-to-media (e.g. in seeking for media support) and citizens-to-developers (e.g. in initiating the open data-driven projects and receiving technical support) (see Fig. 2.2).

Even though the government-centered standpoint is quite ubiquitous in academic literature and generally accepted as a classical one, the civic perspective is becoming increasingly popular both among

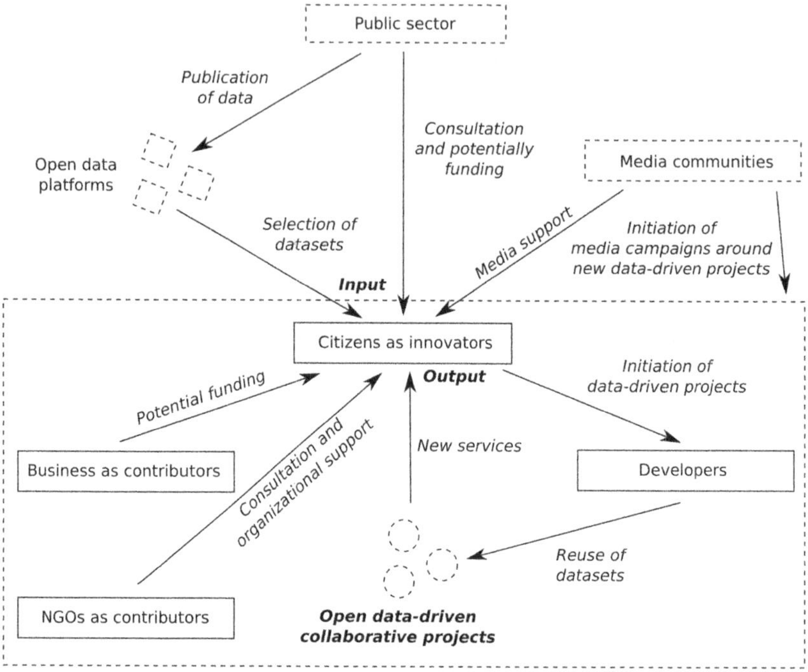

Fig. 2.2 Understanding the crucial role of civic actors in innovating with open data governance. (Source: Own illustration)

practitioners and theorists in the non-governmental sector, especially in various NGOs, open data schools and think tanks (Kassen, 2015a) due to the emergence of new tools to boost civic engagement projects in public sector reforms, especially in the light of the recent diffusion of the open data movement where autonomy and discretion in decision-making could arguably play a really important role in the overall sustainability of independent data-driven digital projects. In this regard, the civic actors could be hypothetically regarded as the most important innovators of the open data concept since they initiate the ideas for various independent open data-driven projects and startups and then seek further support from other stakeholders in the market, not always being motivated by financial reasons.

Research question # 2:
Taking into account the highly collaborative nature of open data, it seems interesting to understand how such new data-driven networking layers as citizens-to-citizens (e.g. independent peer-to-peer open data projects), citizens-to-businesses (e.g. open data-driven e-commerce projects) and, more importantly, citizens-to-NGOs (e.g. the participation of people in various non-governmental initiatives, including those that operate in the open data area), citizens-to-media (e.g. the participation of citizens in the public scrutiny of government agencies through mass media) and citizens-to-developers (e.g. the cooperation of people with developers in launching various open data-driven independent projects, sharing and crowd-sourcing platforms) are manifested in real-life contexts.

3 The Business Perspective: Could Open Data-Driven Governance Provide an Effective Tool of Public-Private Partnerships?

The business perspective is probably the oldest one in the open data area since in the corporate world such dyadic pairs as business-to-business (i.e. classic business relations), business-to-citizens (e.g. the promotion of various e-services for citizens) and business-to-government (e.g. the participation of businesses in public procurement, co-investment in various e-government systems and public-private partnerships) are well-known classic components of the developed digital economy and related terminology had been widely accepted long before the open data government concept emerged, especially in such areas as, for example, e-commerce that deals with huge amount of commercial and corporate data. In this regard, the assimilation of open data-driven processes in public administration in accordance with the long ago established practices in the business sector is regarded as one of the most perspective ways to progress in the area (Ferro & Osella, 2013; Janssen & Zuiderwijk, 2014; Kassen, 2019c; Lakomaa & Kallberg, 2013; Magalhaes et al., 2013; Pantano et al., 2019).

It is especially true if policymakers would focus on the most effective and proven aspects that presumably could be adapted to the needs of public sector reforms, encouraging businesses to participate in the processes and be an active initiator of the transformations. In this respect, it seems interesting to understand from the perspective of business people how open data governance is usually viewed, ideally regardless of cultural contexts. Whether it is considered as a technology-driven platform in public administration, which primary goals are to reduce corruption in government agencies, increase communication between government, businesses and citizens and even provide new horizons for the privatization of some public services in the future, for example, through the notorious mechanism of electronic public procurement (e-procurement) or mere cooperation in the development of shared digital centers or, more importantly, further processing and reuse of government datasets, the business perspective could provide a unique pragmatic view on the perception of open data philosophy as a mutually beneficial enterprise, too. The analysis of such perspectives could help to find answers on how to make the concept economically more sustainable and cost-effective in the future, benefiting both public and private sectors (see Fig. 2.3).

In this regard, in its practical sense, the empirical investigation of the open data movement in its real-life settings and, furthermore, in a cross-country context could also include such components of analysis, for example, as *business-to-NGOs* (e.g. in the cooperation of business with various charity funds and philanthropic activities), *business-to-media* (e.g. in the promotion of various corporate PR-activities and contacts with electronic and print media) and *business-to-developers* (especially, if business regards contacts with developers as one of the main components of success in data-driven markets, e.g., in embracing the philosophy of big data, developing open data-driven information systems and related networks). In general, taking into account that in many developed countries business communities regard open data as a certain continuation of big data-driven e-commerce philosophy in public administration (Ciancarini et al., 2016), they sometimes indirectly participate in promoting open data-driven projects that could benefit their business, mostly through funding and knowledge sharing and consider it as a public value (Magalhaes & Roseira, 2017).

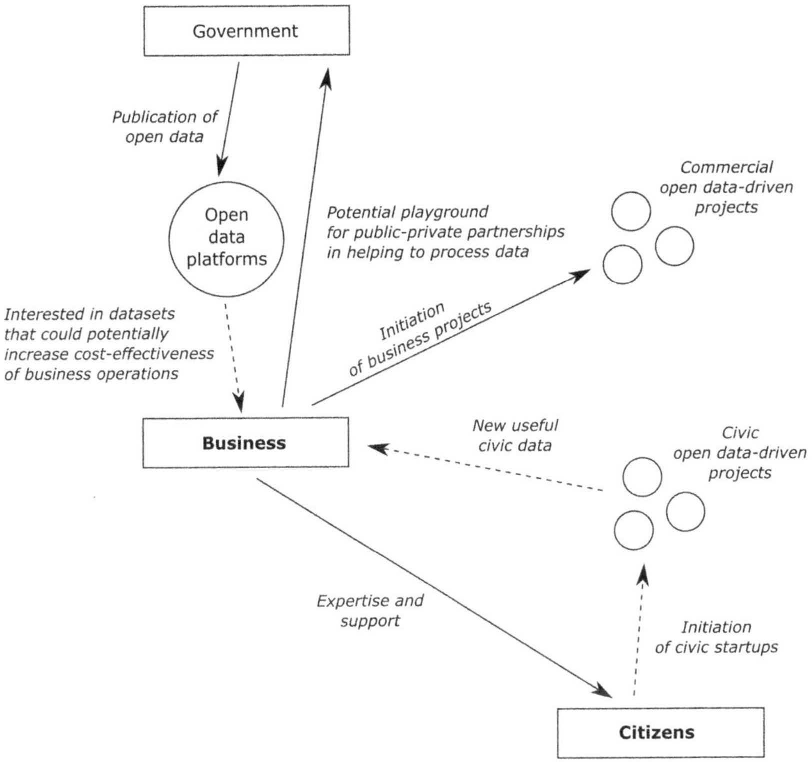

Fig. 2.3 Understanding the role of the private sector in advancing open data as a promising source for business innovations. (Source: Own illustration)

Research question # 3:
Open data governance could hypothetically reduce the presence of the public sector in the market, leaving more space for the private sector to expand and prosper due to small and more effective technology-driven government and related public service ecosystems it has to operate in. In this respect, it will be interesting to test these assumptions and understand what the primary motivations of the business actors are to join the open data movement and advance the concept as a collaborative phenomenon, especially in cooperating and partnering with various stakeholders.

4 Developers as Actors: Is It Only All About Procurement of Open Data-Driven Innovations?

The professional attitude of developers toward the open data governance philosophy is probably one of the most interesting areas to study, since their role in the development of various open data-driven public administration systems is usually widely overlooked and often merely considered as a technological, that is, a passive one, which activity in the area is deemed to be based predominantly on demands from government agencies, for example, in designing the user interface or providing technological and telecommunication services and consultations to the public sector. This explains a gap in the academic research on the topic (Brunswicker et al., 2018; Dogan et al., 2021; Ferro et al., 2013; Kassen, 2018), where the main focus of investigation is usually on governments, citizens and businesses and less often on developers as key stakeholders of the projects in the area, even though Information and Communication Technology (ICT) experts play a very important role in maintaining organizational, technological and even financial sustainability of the projects (Ojo et al., 2015). Moreover, in a classical e-government paradigm, due to the fact that many services of developers are often procured by the public sector, which usually considers itself as an initiator for a lion's share of technology-driven innovations in government, they are simply regarded as subjects that indirectly represent the former and, therefore, their political role in the processes should ostensibly be deemed as almost non-existent. In this regard, in order to see a better picture about the multidimensional roles of developers as actors of the open data movement in a more holistic manner, it is necessary to distinguish independent developers from those who are actively engaged in the development of related open data projects in both public and private sectors of the economy and thus could hypothetically be biased by contractual obligations (see Fig. 2.4).

With the development of more independent models in building open data systems, the situation could dramatically change, especially when the social role and even political vision of specialists and experts in the area toward open data-driven public sector reforms and related projects could begin to play more important and, sometimes, even crucial role in the promotion of various independent data-driven civic engagement and e-participation platforms. After all, networking among specialists in the area could be a core of the success in many independent civic projects and

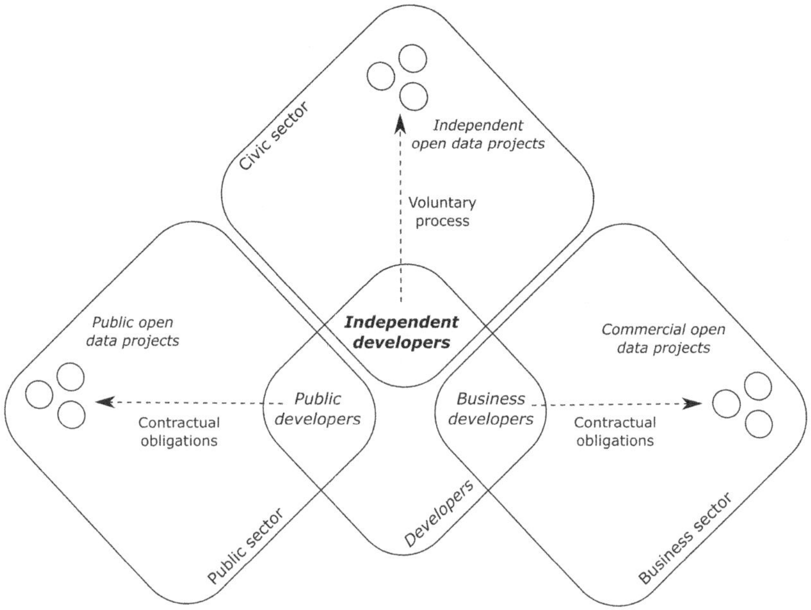

Fig. 2.4 Understanding the multidimensional roles of developers as actors of the open data movement. (Source: Own illustration)

startups. In this respect, it is important to understand the true motives that drive the genuine interest of professional communities in such open data-driven projects. In particular, it seems interesting to understand what open data movement really is from the perspective of developers, what kind of projects they prefer to create, what operating platforms they regard most promising to promote, systematic challenges they face and so on. Another important aspect of analysis could be the investigation of key socioeconomic factors that drive independent developers in the area to reuse open data: whether it is as an excellent opportunity to apply available knowledge or maybe test various venturous projects and initiatives in the area on both paid and unpaid basis.

The key hypothesis argued in this connection is that when it comes to the realization of open data-driven projects, where a developer could play a more important socioeconomic role in the overall sustainability of such projects, he or she should be regarded as an independent actor of open data systems, especially in launching various civic engagement projects, which are often initiated and supported by individual enthusiasts themselves with or without support from public or private sectors. It could be easily noticeable at local levels, where, for example, many open data-driven projects exist only due to the philanthropic and volunteer support of local developers and technically savvy citizens. In this respect, it is also interesting to understand how, according to the developer's perspective, the open data concept is understood in relationship to the study of such binary pairs as, for example, *developers-to-government* (e.g. in helping the public sector to reform through various open data-driven government systems and platforms), *developers-to-business* (e.g. in developing corporate open data-driven e-commerce systems), *developers-to-citizens* (e.g. in helping citizens to launch various independent open data-driven or wiki projects), *developers-to-NGOs* (e.g. in cooperating with non-governmental players, universities and think tanks, developers could share expertise and knowledge in the area), *developers-to-developers* (e.g. in collaborating through various online and offline forums and hackathons, webinars and master classes) and *developers-to-media* (e.g. in explaining the latest developments in the open data area, participation in public awareness projects, public interviews and providing expert opinions).

Research question # 4:
The experts in the area could presumably regard open data governance as a technological platform, which embraces the best developments in the ICT sector in order to make public administration and political communication with citizens and businesses better and more transparent or they may simply consider their role as a technological bridge between other key players, which, unfortunately, is often forgotten by many policymakers and widely regarded as an insignificant part of the open data concept. Moreover, many developers could also be dependent players who mostly serve the government in the technological reformation of public administration and, therefore, in this case, they should be regarded as part of the public sector itself. Such assumptions could also be tested in a real life context.

5 The Perspective of the Non-governmental Sector: Could Open Data Serve the Promise of Digital Democracy?

The perspective of the non-governmental sector is, probably, the most non-standard one in open data governance philosophy, since it is not popular everywhere and its presence and role might range from virtually non-existent or insignificant in totalitarian and authoritarian contexts to highly important and often crucial one in developed democracies and open societies. For example, as in case with traditional e-government technologies, authoritarian governments could try to develop and sometimes even formally succeed in promoting open data platforms, usually using the standard e-government procedures and tactics yet in a very top-down and strictly directive manner with almost no participation of the non-governmental and civic sectors in the process in an attempt to disguise a truly oppressive nature of political regimes there. Likewise, the same as in case with developers, the non-governmental sector could predictably consider its role as a mere proxy or intermediary between society and government. In this respect, it seems interesting to shed light on its main goal and function as a presumably important third-party actor in defending the interests of civil society in communication with public agencies. Furthermore, it is important to understand how the core conceptual values of open data governance as a political concept is diffusing among various think tanks, academia, advocate organizations and groups of interests in the area and whether it is regarded as an effective platform to reform not only public administration but, more importantly, social and political systems by *indirectly* helping various stakeholders of the open data movement with various social programs, hypothetically changing the very core of communication and even political culture through various technology-driven projects and initiatives (see Fig. 2.5).

Traditionally, the non-governmental sector includes such components as public activists, universities, research centers and think tanks, philanthropic organizations, independent development funds, private science foundations, charity funds, club of interests, lobbying groups and so on. Often the role of non-governmental organizations in the digital arena is associated with its representative function as a civil society institution to ensure public scrutiny of public administration systems for citizens (Sein, 2011; Vosloo & Van Belle, 2005) or as a watchdog in promoting various anti-corruption platforms in the e-government area and other projects

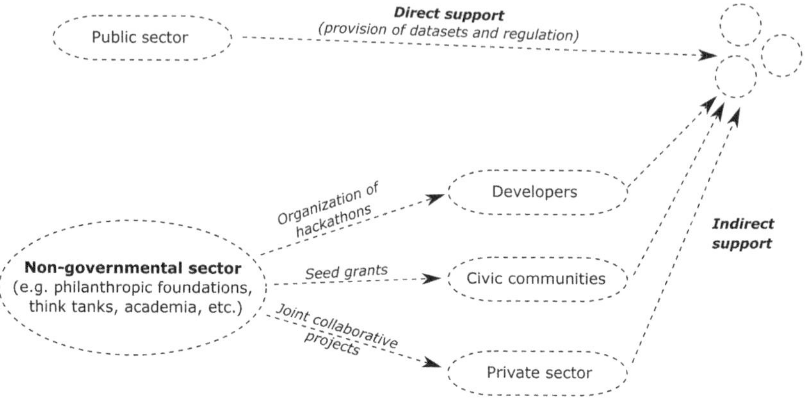

Fig. 2.5 Understanding the indirect role of the non-governmental organizations in advancing open data as a collaborative movement. (Source: Own illustration)

(Welch & Pandey, 2007), especially of those that, for example, are aimed at helping the public sector to reduce the level of digital divide, which in a dyadic manner could be described as *NGOs-to-e-government transactions*. However, there is no research that would study the topic in the context of open data phenomena. In this regard, in order to fill the gap in academic literature, it would be interesting to look at the role of non-governmental actors in cooperation also with other stakeholders, especially in the private and civic sectors of economy in advancing the concept of open data governance as a truly collaborative phenomenon.

Research question # 5:
Other promising venues for empirical analysis could be found in researching such interactions as NGOs-to-citizens and NGOs-to-developers transactions, for example, in helping to engage citizens and developers; NGOs-to-business interactions, especially in an attempt to understand the fundraising activity and diversification of funding options for various open data-driven e-commerce projects; NGOs-to-media, for example, in identifying the invisible mechanism of open data-driven collaboration that could relate, for example, to the investigation of cooperative media coverage of various civic engagement programs in the area, public awareness programs and supporting research and related curricula in various institutions of tertiary education.

6 Media as an Actor: Understanding the Roles of Journalistic Communities in Raising Public Awareness on Open Data Governance?

The media perspective is the youngest one in the theoretical paradigm of the open data concept among both practitioners and academics. One of the promises of the media approach is that journalistic communities presumably could affect such aspects of political communication as, for example, transparency of governance and trust in government (Gün et al., 2021; Janssen et al., 2012; Kassen, 2019d) in covering related government reforms, including those that are associated with the adoption of data-driven journalism (Böhm et al., 2010; Gray et al., 2012; Kassen, 2015b; Stoneman, 2015) and other technology-driven public sector projects in the area. With the rapid developments in open data technologies, it is interesting to know about the role of media in the processes, too, paying special attention to various public awareness programs in the open data area as well as to find out whether local journalistic communities regard the concept as an additional effective tool to maintain public relations with government agents and officials. The media perspective might prove to be the most eye-opening in the open data research, taking into account the lack of virtually any previous works in this area of investigation and paradoxical abundance of affluent empirical evidence of participation from the part of journalists in the processes. After all, by resorting to new information technologies in various political campaigns, especially in social media (Brajawidagda & Chatfield, 2014; Jalali et al., 2017; Janssen & Zuiderwijk, 2014), an example of which are Barack Obama's 2008 and 2012 presidential campaigns (Cogburn & Espinoza-Vasquez, 2011; Hong & Nadler, 2012), public sector reforms and social network-focused political activities and political technologists have dramatically increased the mass media's interest to various open data projects, in both public and private sectors of economy (see Fig. 2.6).

In this regard, such new components of open data governance paradigm as media-to-government (e.g. in the contacts of mass media with government through information and communication technologies), media-to-citizens (e.g. in the promotion of interactive television and social media channels), media-to-business (e.g. in the participation of mass media in covering various open data-driven e-commerce systems), media-to-NGOs (e.g. in helping the non-governmental sector to propagate the public value of open data and the promise of civic engagement among citizens), media-to-developers (e.g. in helping to engage developers in various open data-driven projects) and so on—all should be considered as an

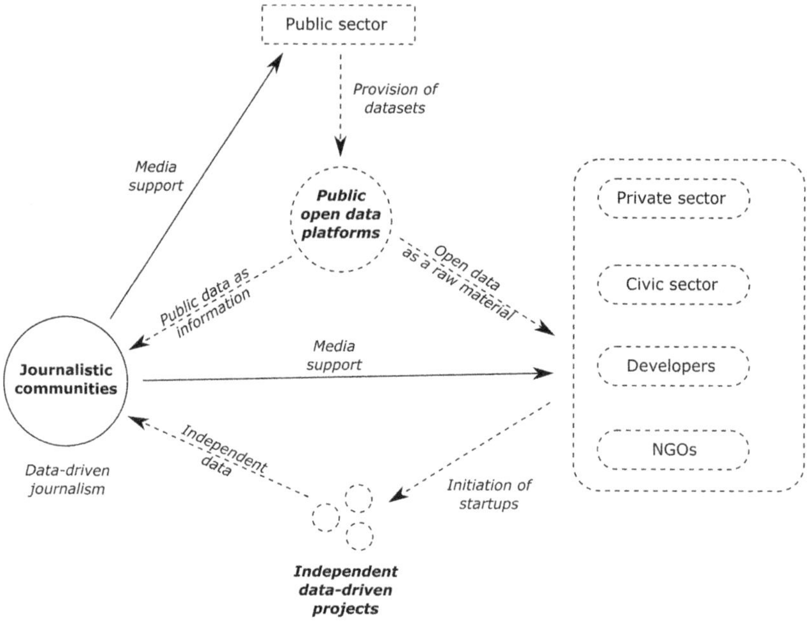

Fig. 2.6 Understanding the supportive role of media communities in advancing the open data movement. (Source: Own illustration)

intrinsic part of modern information society. The fact that many open data-driven platforms maintain various accounts in social networks, such as Facebook, Twitter, Instagram, Google Plus and others, says that contacts with people and, more importantly, with media, which regards itself often as a medium between citizenry and government, should be valued today in modern societies as never before and any research in this direction could prove to be really promising.

Research question # 6:
Taking into account the fact that media are beginning to play an ever-increasing role in raising public awareness about open data as a collaborative movement, it is interesting to understand how related journalistic communities use open data in their everyday activities and what motivates them to join the movement and participate in various media coverage campaigns initiated by both public and private sectors of economy.

7 Conclusion

The analysis of academic literature related to the contributing roles on each of the open data actors, both public and private, clearly demonstrates that the open data movement is an intrinsically multidimensional and networking phenomenon, especially when it comes to the reuse and reapplication of various government datasets in third-party data-driven projects, startups and initiatives. These projects are usually initiated by civic, non-governmental and business stakeholders to propose new innovative services or information products, the demand on which could not be necessarily met within traditional e-government, e-participation or e-commerce ecosystems. Collaboration, co-creation and participation are key pillars that help to promote the movement since open data developers are often driven purely by philanthropic motivations, whereas media and academic communities play an important role as infomediaries in the process. In this regard, it is interesting to test all these networking interactions in a more holistic manner by resorting to the empirical case study analysis from the perspectives of two nations in Nordic Europe that are currently leading globally the open data movement.

References

Böhm, C., Naumann, F., Freitag, M., George, S., Höfler, N., Köppelmann, M., ... & Schmidt, T. (2010, September). Linking open government data: What journalists wish they had known. In Proceedings of the 6th international conference on semantic systems, pp. 1–4.

Boughzala, I., Janssen, M., & Assar, S. (2015). E-government 2.0: Back to reality, a 2.0 application to vet. In *Case studies in e-Government 2.0* (pp. 1–14). Springer.

Brajawidagda, U., & Chatfield, A. T. (2014). *Roles of social media in open data environments: A case study of the 2014 Indonesian presidential election voting results.*

Brunswicker, S., Majchrzak, A., Almirall, E., & Tee, R. (2018). Cocreating value from open data: From incentivizing developers to inducing cocreation in open data innovation ecosystems. *World Scientific Book Chapters, 3*, 141–162.

Ciancarini, P., Poggi, F., & Russo, D. (2016, March). Big data quality: A roadmap for open data. In 2016 IEEE second international conference on big data computing service and applications (BigDataService) (pp. 210–215). IEEE.

Ciociola, C., & Reggi, L. (2015). *A Scuola di OpenCoesione: From open data to civic engagement* (p. 26). Open Data as Open Educational Resources.

Cogburn, D. L., & Espinoza-Vasquez, F. K. (2011). From networked nominee to networked nation: Examining the impact of Web 2.0 and social media on political participation and civic engagement in the 2008 Obama campaign. *Journal of Political Marketing, 10*(1–2), 189–213.

Conradie, P., & Choenni, S. (2014). On the barriers for local government releasing open data. *Government Information Quarterly, 31*, 10–17.

Dogan, G., Taskin, Z., & Aydinoglu, A. U. (2021). Research data management in Turkey: A survey to build an effective national data repository. *IFLA Journal, 47*(1), 51–64.

Ferro, E., Caroleo, B., Leo, M., Osella, M., & Pautasso, E. (2013). The role of ICT in smart cities governance. In *Proceedings of 13th international conference for E-democracy and open government* (pp. 133–145). Donau-Universität Krems.

Ferro, E., & Osella, M. (2013, April). Eight business model archetypes for PSI re-use. In Open Data on the Web Workshop. Google Campus, Shoreditch, London. https://www.w3.org/2013/04/odw/odw13_submission_27.pdf

Fusi, F., & Feeney, M. K. (2020). Data sharing in small and medium US cities: The role of community characteristics. *Public Administration, 98*(4), 922–940.

Gonzálvez-Gallego, N., & Nieto-Torrejón, L. (2021). Can open data increase younger generations' trust in democratic institutions? A study in the European Union. *Plos one, 16*(1), e0244994.

Gray, J., Chambers, L., & Bounegru, L. (2012). *The data journalism handbook: How journalists can use data to improve the news.* O'Reilly Media, Inc.

Gün, A., Pak, B., & Demir, Y. (2021). Responding to the urban transformation challenges in Turkey: A participatory design model for Istanbul. *International Journal of Urban Sustainable Development, 13*(1), 32–55.

Helbig, N., Gil-García, J. R., & Ferro, E. (2009). Understanding the complexity of electronic government: Implications from the digital divide literature. *Government Information Quarterly, 26*(1), 89–97.

Hong, S., & Nadler, D. (2012). Which candidates do the public discuss online in an election campaign?: The use of social media by 2012 presidential candidates and its impact on candidate salience. *Government Information Quarterly, 29*(4), 455–461.

Jalali, J., Mohammad, S., & Park, H. W. (2017). Conversations about open data on Twitter. *International Journal of Contents, 13*(1).

Janssen, M., Charalabidis, Y., & Zuiderwijk, A. (2012). Benefits, adoption barriers and myths of open data and open government. *Information Systems Management, 29*(4), 258–268.

Janssen, M., & Zuiderwijk, A. (2014). Infomediary business models for connecting open data providers and users. *Social Science Computer Review, 32*(5), 694–711.

Joseph, B. K. (2019). Blockchain for open data – Exploring conceptual underpinnings and practice. In *Governance models for creating public value in open data initiatives* (pp. 161–175). Springer.

Kassen, M. (2013). A promising phenomenon of open data: A case study of the Chicago open data project. *Government Information Quarterly, 30*(4), 508–513.

Kassen, M. (2015a). *Understanding systems of e-Government: e-Federalism and e-Centralism in the United States and Kazakhstan*. Lexington Books.

Kassen, M. (2015b). E-Government in the United States: The federal model of implementation. In *Understanding systems of e-Government* (pp. 27–43). Rowman & Littlefield.

Kassen, M. (2016). *E-government in Kazakhstan: A case study of multidimensional phenomena*. Routledge.

Kassen, M. (2018). Open data and its intermediaries: A cross-country perspective on participatory movement among independent developers. *Knowledge Management Research & Practice, 16*(3), 327–342.

Kassen, M. (2019a). *Open data politics: A case study on Estonia and Kazakhstan*. Springer.

Kassen, M. (2019b). Open data and e-government–related or competing ecosystems: A paradox of open government and promise of civic engagement in Estonia. *Information Technology for Development, 25*(3), 552–578.

Kassen, M. (2019c). Open data politics in Estonia: Advancing open Government in the context of Ubiquitous digital state. In *Open Data Politics* (pp. 37–67). Springer.

Kassen, M. (2019d). Open data politics in Kazakhstan: Understanding a tentative advance of civic engagement in a transitional society. In *Open data politics* (pp. 69–98). Springer.

Kassen, M. (2020). Open data and its peers: Understanding promising harbingers from Nordic Europe. *Aslib Journal of Information Management, 72*(5), 765–785.

Kassen, M. (2021a). Blockchain and e-government innovation: Automation of public information processes. *Information Systems, 103*, 101862.

Kassen, M. (2021b). Understanding decentralized civic engagement: Focus on peer-to-peer and blockchain-driven perspectives on e-participation. *Technology in Society, 66*, 101650.

Kitchin, R. (2014). *The data revolution: Big data, open data, data infrastructures and their consequences*. Sage.

Lakomaa, E., & Kallberg, J. (2013). Open data as a foundation for innovation: The enabling effect of free public sector information for entrepreneurs. *IEEE Access, 1*, 558–563.

Magalhaes, G., & Roseira, C. (2017). Open government data and the private sector: An empirical view on business models and value creation. *Government Information Quarterly, 37*, 101248.

Magalhaes, G., Roseira, C., & Strover, S. (2013, October). Open government data intermediaries: A terminology framework. In proceedings of the 7th international conference on theory and practice of electronic governance, pp. 330–333.

Meijer, A., & Potjer, S. (2018). Citizen-generated open data: An explorative analysis of 25 cases. *Government Information Quarterly, 35*(4), 613–621.
Meijer, R., Conradie, P., & Choenni, S. (2014). Reconciling contradictions of open data regarding transparency, privacy, security and trust. *Journal of Theoretical and Applied Electronic Commerce Research, 9*(3), 32–44.
O'Hara, K. (2012, June). Transparency, open data and trust in government: Shaping the infosphere. In Proceedings of the 4th annual ACM web science conference, pp. 223–232.
Ojo, A., Curry, E., & Zeleti, F. A. (2015, January). A tale of open data innovations in five smart cities. In 2015 48th Hawaii international conference on system sciences (pp. 2326–2335). IEEE.
Ojo, A., & Janssen, M. (2013, June). Aligning core stakeholders' perspectives and issues in the open government data community. In Proceedings of the 14th annual international conference on digital government research, pp. 293–294.
Ojo, A., & Millard, J. (Eds.). (2017). *Government 3.0 – Next generation government technology infrastructure and services: Roadmaps.* Enabling Technologies & Challenges (Vol. 32). Springer.
Pantano, E., Priporas, C. V., Stylos, N., & Dennis, C. (2019). Facilitating tourists' decision making through open data analyses: A novel recommender system. *Tourism Management Perspectives, 31*, 323–331.
Pereira, G. V., Macadar, M. A., Luciano, E. M., & Testa, M. G. (2017). Delivering public value through open government data initiatives in a Smart City context. *Information Systems Frontiers, 19*(2), 213–229.
Puussaar, A., Johnson, I. G., Montague, K., James, P., & Wright, P. (2018). Making open data work for civic advocacy. Proceedings of the ACM on Human-Computer Interaction, 2(CSCW), pp. 1–20.
Ricker, B., Cinnamon, J., & Dierwechter, Y. (2020). *When open data and data activism meet: An analysis of civic participation in Cape Town, South Africa.* The Canadian Geographer/Le Géographe canadien.
Sein, M. K. (2011). The "I" between G and C: E-government intermediaries in developing countries. *The Electronic Journal of Information Systems in Developing Countries, 48*(1), 1–14.
Sieber, R. E., & Johnson, P. A. (2015). Civic open data at a crossroads: Dominant models and current challenges. *Government Information Quarterly, 32*(3), 308–315.
Stoneman, J. (2015). *Does open data need journalism?* University of Oxford.
Susha, I., & Gil-Garcia, J. R. (2019, January). A collaborative governance approach to partnerships addressing public problems with private data. In Proceedings of the 52nd Hawaii international conference on system sciences.
Tinholt, D. (2013). *The open data economy: Unlocking economic value by opening government and public data.* Capgemini Consulting.
Vosloo, S., & Van Belle, J. P. (2005, August). E-government and the E-readiness of Non-Profit Organisations in the Western Cape. In The 2nd annual confer-

ence of the Community Informatics Research Network (CIRN), South Africa, pp. 24–26.

Weerakkody, V., Irani, Z., Kapoor, K., Sivarajah, U., & Dwivedi, Y. K. (2017). Open data and its usability: An empirical view from the Citizen's perspective. *Information Systems Frontiers, 19*(2), 285–300.

Weerakkody, V., Kapoor, K., Balta, M. E., Irani, Z., & Dwivedi, Y. K. (2017). Factors influencing user acceptance of public sector big open data. *Production Planning & Control, 28*(11–12), 891–905.

Welch, E. W., & Pandey, S. K. (2007). E-government and bureaucracy: Toward a better understanding of intranet implementation and its effect on red tape. *Journal of Public Administration Research and Theory, 17*(3), 379–404.

Yavuz, M. F., & Saiti, B. (2020). Eliminating trust issues in Islamic interbank monetary system by Blockchain technology. In *Impact of Financial Technology (FinTech) on Islamic finance and financial stability* (pp. 1–11). IGI Global.

Yli-Huumo, J., Päivärinta, T., Rinne, J., & Smolander, K. (2018). Suomi.fi – Towards government 3.0 with a national service platform. In *International conference on electronic government* (pp. 3–14). Springer.

CHAPTER 3

Understanding the Perspectives of Peer-To-Peer Actors in the Open Data Movement

Abstract The peer-to-peer perspective is probably the most promising area in researching the collaborative nature of the open data governance paradigm since it could help to better understand the internal driving forces for various actors to join the movement at various horizontal institutional levels. Peers' outlook on open data reveals interesting communicative venues in the development of *closed* or *internal* open data-driven government networks, related autonomous civic engagement projects or even independent corporate business-to-business e-commerce projects in the area. It is necessary to note that this direction of open data studies is the least developed in academic and professional literature, taking into account a closed nature of such transactions and anonymity or confidential character of related networking platforms.

Keywords Open data • Peers • Peer-to-peer • Government peers • Business peers • Civic peers

M. Kassen, *Open Data Governance and Its Actors*, Studies in National Governance and Emerging Technologies,
https://doi.org/10.1007/978-3-030-92065-4_3

1 Government Peer-to-Peer Interactions: Is There Any Inter-institutional Balance Between Centralization in Generating Datasets and Provision of Proper Information Security?

The government peer-to-peer or government-to-government transactions are part of the oldest, most conservative and bureaucratic dimensions in technology-driven political communication since the phenomenon of open data government has appeared first as an administrative concept in public institutions, which as a trend could be traced globally (Abu-Shanab, 2015; Akcagündüz, 2021; Geiger & von Lucke, 2011; Höchtl & Reichstädter, 2011; Kassen, 2014; Kassen, 2019; Mergel et al., 2018; Sadiq & Indulska, 2017; Yıldırım & Bostancı, 2021). In this regard, the generation of a huge stockpile of government datasets within the existing ecosystem of e-government platforms in a really wide range of activities allows the public sector to select, pre-process and publish some of the datasets in a transparent manner. Taking into account the fact that public sector has always been a prime initiator and promoter of all digital government initiatives in all its extremely diverse forms and technological versions (Feeney & Welch, 2016), including in open data governance, the centralized generation of related government datasets at both national and local levels seems to be the most promising venue where one should look first at, especially if it is necessary to demonstrate the public value of such peer-to-peer transactions. For example, in order to better realize the intrinsically enclosed nature of peer-to-peer transactions in government, especially in regard to the generation and use of official information, one should imagine a networking framework, which is, on the one hand, extremely secretive yet very effective organizationally, encompassing a wide range of internal communication links and inter-institutional transactions between various public agencies at both national and local levels as well as diplomatic and special services that use the same databases in a strictly centralized and systematic manner. Likewise, the related collaboration of government agencies could contribute to a more effective centralized generation of new datasets, some of which may later be published as open data (see Fig. 3.1).

At the first glance, the idea might seem to be as an additional useful tool to reform public administration and internal administrative processes (Kandhro et al., 2013), especially if policymakers pursue the only primary

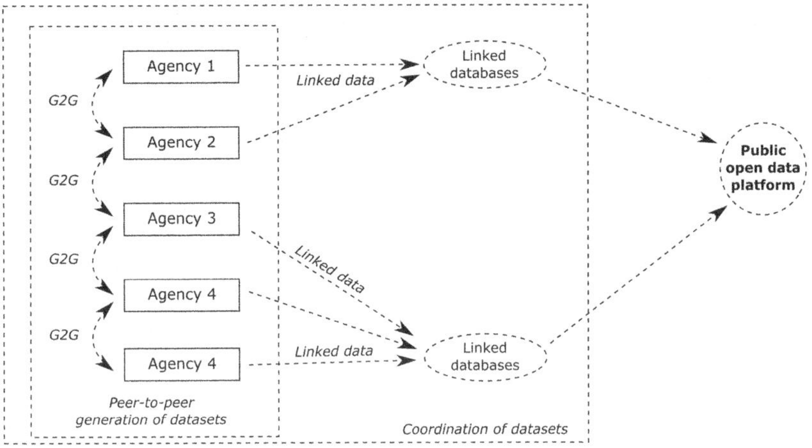

Fig. 3.1 The invisible mechanism of government-to-government generation of raw and linked open data. (Source: Own illustration)

goal of such transactions, that is to make government more efficient and cost-effective (Obama, 2009). However, it is necessary to understand that the informatization of public sector reforms is a main driver in the advancement of such systems today in public administration itself while new promising networks of data-driven inter-institutional communication channels that could be promoted between government agencies themselves could be just a mere side effect of such interactions, which not necessarily leads to greater transparency without intensive demands for public scrutiny in the area. After all, even though the government-to-government communication might be sometimes open for public scrutiny, especially if social media platforms are used in such interactions, the government networks themselves are usually closed and in some cases even encrypted to ensure confidentiality of transactions whether it is related to governmental communication channels between domestic national and local public bodies or between diplomatic or intelligence services abroad and their administrative or regional branches.

In this regard, along with obvious advantages such as *centralized generation and publication of government datasets*, this internal administrative institution could present some challenges, too. After all, for hackers and intruders, it would be much more promising to tamper centralized datasets rather than decentralized. Taking into account the intricate structure of government networks and security lapses always existing in information systems, data and information leaks do occasionally happen due to external malicious activity of ICT hackers or, sometimes, due to internal deliberate activities of whistleblowers. The closed institutional environment of internal government transactions, especially between different bureaucratic and administrative levels, may not necessarily guarantee immunity from information leaks and tampering of such centralized datasets. Likewise, the closed nature of internal peer-to-peer transactions might give birth to various conspiracy theories and legends in information society and the topic itself in e-government and, more importantly, in open data areas could be considered as one of the most politically debated and controversial among both policymakers and practitioners. *Therefore, any empirical research from a cross-country perspective could provide new much needed material for hypothesis testing and further studies in this area. In this regard, the matter of centralization and decentralization in the generalization of government datasets could academically present the most interesting venue for practical investigation in the area in both of the studied countries, taking into account the unitary nature of e-government settings there.*

2 Civic Peer-to-Peer Interactions: Is Civic Engagement a Sustainable Mechanism to Advance the Open Data Movement?

The study of civic peer-to-peer or citizen-to-citizen transactions in regard to open data is another promising actor's perspective to research in the area, especially in a political sense since such independent networks could help to revolutionize the concept into a truly collaborative platform, especially for cooperation between members of civil society itself. The main public value of civic peer-to-peer interactions might be found in the opportunities that the reuse of open data-driven communication could offer for local communities and urban neighborhoods in improving their lives and democratic dialogues with government. Taking into account the collaborative nature of open data phenomena, the related cooperation between civic stakeholders hypothetically allows resorting to the power of crowd-sourcing and funding, especially if new proposed open data-driven services benefit the whole community or neighborhood. Technically savvy citizens could self-organize and advance the open data concept as a focal point for various initiatives at local levels, harnessing organizational, technological and financial help from other peers, who are united by a single goal to make local communities better through civic open data-driven self-governance (see Fig. 3.2).

In this respect, such technological platforms as wiki-based (Frank & Zander, 2017), social media peer-to-peer (Gasco et al., 2020) or

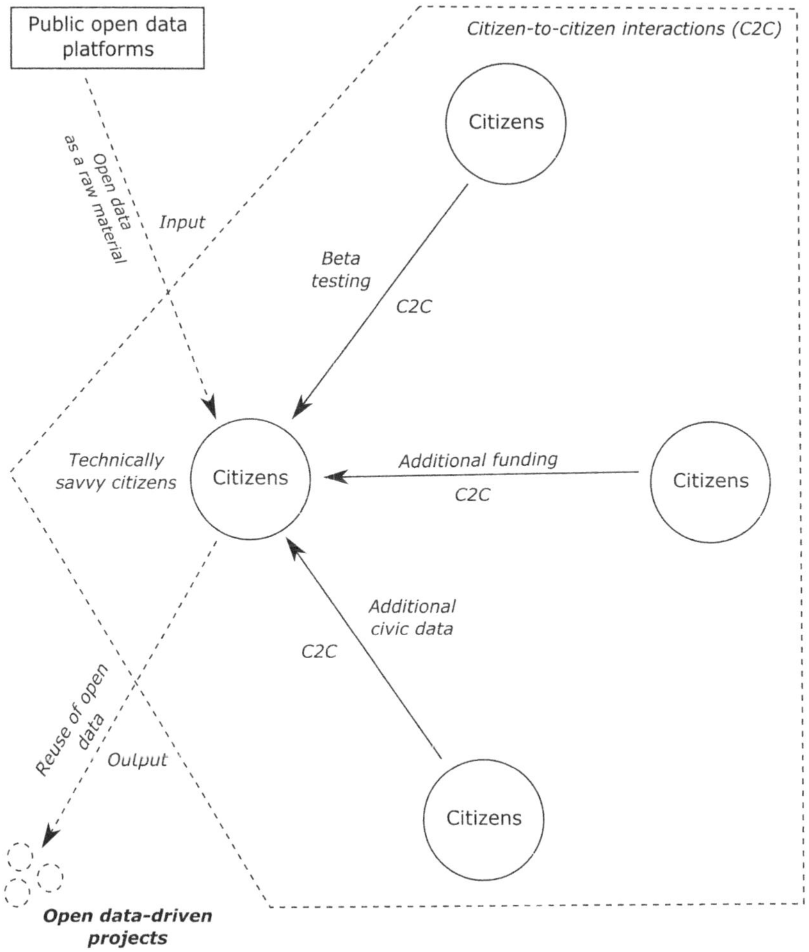

Fig. 3.2 Understanding the collaborative nature of civic peer-to-peer open data-driven phenomena. (Source: Own illustration)

developers peer-to-peer open data-driven projects (Kassen, 2020), that harness the potential of civic engagement in furthering independent open data-driven projects, could regard public agencies only as a certain reliable resource of datasets, presumably valuing more networking opportunities that arise in the environment of participation among, first of all, citizens than official interactions with government agencies itself, prioritizing more civic engagement than top-down character of communication in such transactions and iconizing the power of collective knowledge and wisdom that arguably should be leveraged in such projects to achieve any success in this direction. As such, the author of the research argues that the main public value of open data governance from that civic peer-to-peer perspective is not so much that it gives the opportunity, for example, to reform public administration and battle red-tape, it is that it opens new niches to advance the participatory nature of democracy in its original direct form, revolutionizing even the old good idea of e-government itself, making it more politically oriented and focused on collaboration and participation rather than on transparency and accountability. In this regard, it is necessary to praise the importance of Barack Obama's wise outreaching policy to promote both nationally and globally the idea of open data government as a certain type of e-governance that was conceptually reincarnated in a new more politically oriented manner. *All these assumptions need to be analyzed in detail through the prism of real examples of civic open data-driven projects and initiatives, especially in understanding how such projects are initiated, developed and promoted among various civic communities and what kind of assistance is offered by other civic peers to those technically savvy citizens who started the movement at local levels in both countries.*

3 Business Peer-to-Peer Interactions: Could It Present Any Successful Mechanism for Sustainable Open Data-Driven Business?

The business peer-to-peer or business-to-business open data-driven transactions are usually regarded by business communities as an important component in various digital public projects, especially in those that promote the cooperation between government, investors and business people equally at global, national and local levels of governance (Joseph, 2019; Kassen, 2016; Kubler et al., 2018; Pereira et al., 2015; Sebetci, 2015;

Sebetci & Aksu, 2014). The participation of various transnational telecommunication and consulting companies such as IBM, Oracle, Deloitte, Microsoft and many others in the development of an extremely wide range of digital solutions for the public sector in many nations, whether it is in developing e-commerce, e-government, e-voting or e-procurement platforms, is a good example of the general trend in the area. In this regard, it seems interesting to study in detail various platforms designed to draw together businesses, for example, at national levels with their counterparts in government or even with potential partners at the global level. Moreover, such cooperation could be regarded by government agencies as a very effective way to harness the promise of the open data governance concept to boost cooperation between businesses themselves in a true capitalist economy. Therefore, even though the public sector indirectly could support open data-driven projects through grants and investments, the private and, especially, corporate sector should be intrinsically interested in the successful *self-sustainable development* of such projects. For example, business communities may support mutually beneficial public initiatives in the area, especially those that could help them to be more efficient in communicating with each other through digital platforms. Another example of peer-to-peer cooperation in the sector could be a plethora of various corporate open data-driven projects aimed at providing consultative, juridical, informational and even media assistance to small- and middle-sized businesses in launching various startups in the area, dealing with government agencies or entering foreign markets. The potential of such types of communication could be really huge and unlimited in a market economy (see Fig. 3.3).

In general, taking into account the enclosed nature of public administration, all these theoretical assumptions about peer-to-peer transactions in an intrinsically internal institutional environment could remain merely tentative and speculative in theory without real empirical evidence and demonstrative examples. In this regard, in investigating the actual cases of open data-driven projects in government, all the more in a cross-country perspective, it could be interesting to identify and analyze the most promising networking aspects of such peer-to-peer communication in the business sector. It is necessary to pay special attention to the internal sustainable mechanisms of related peer-to-peer platforms in each of the cases and understand to what extent, public agencies, citizens and business entities are ready to use the principle, for example, in resorting to the centralization of all internal government information systems and, more

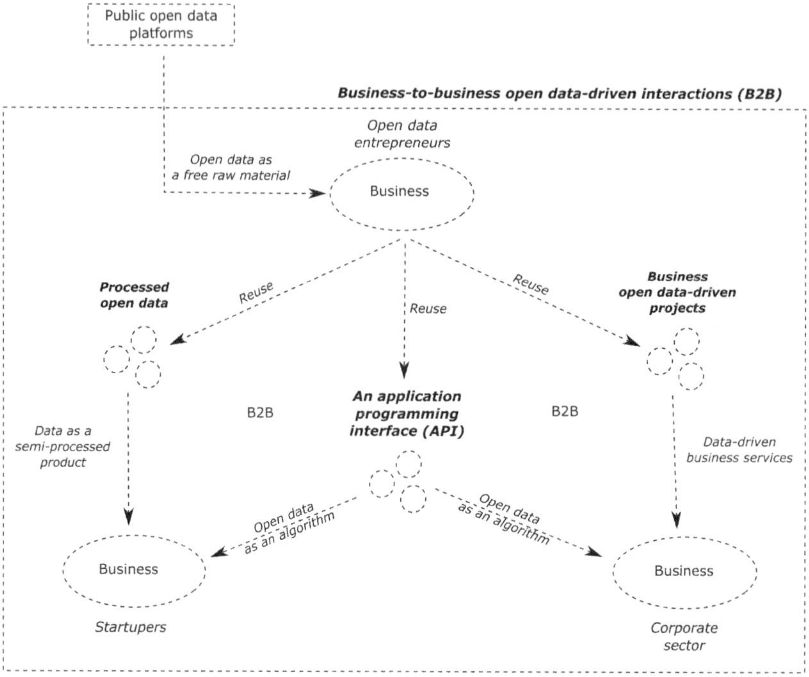

Fig. 3.3 The promise of business peer-to-peer open data-driven interactions. (Source: Own illustration)

importantly, adhering to the centralized models in generating government datasets, which later may be published as open data, offering new information for various business open data-driven projects or proposing promising models to advance various sustainable commercial activities in the area when various business middlepersons or intermediaries could help other open data entrepreneurs to launch their own solutions. *In this regard, the content analysis of official digital government and, more importantly, civic and business open data-driven platforms could help to comprehend the invisible mechanics of related peer-to-peer networking. All these assumptions need to be examined in different socioeconomic contexts, especially in such developed capitalistic and corporate economies as Finland and Sweden.*

4 Open Data Governance from the Perspectives of Different Institutional Levels

All previous above-mentioned individual actor perspectives and related dyadic pairs between peers and other players of digital communication could be regarded as part of open data philosophy at *national and local levels*, which also could hypothetically include other dimensions of technology-driven interactions, especially institutional ones. In this respect, the theoretical investigation presented by the author in this section as, first of all, an institutional analysis is aimed at understanding the key networking aspects of open data-driven communication among various stakeholders at different institutional levels such as *vertical* (e.g. between national and local authorities, businesses and citizens), *horizontal* (e.g. between peers at both national and local levels) and even *international levels* of transactions (e.g. between national and regional government authorities, such as, in the European Union, Latin America and other parts of the world and, furthermore, even with global open data agencies, for instance, the United Nations Public Administration Network, the Open Government Partnership and the World Bank).

4.1 The National and Local Levels of Government Interactions: Do Different Institutional Aspects Really Matter in Promoting Open Data Governance?

The national-local levels of open data-driven transactions are tentatively defined in this research as *invisible vertical digital communication channels* that connect various actors at different institutional echelons of power such as those that may exist in related digital e-government ecosystems between representatives of central and provincial public agencies, local business and civic communities and so on. Institutionally, this type of vertical national local or local-national open data transactions tends to be more popular in countries with federal structures of public administration such as the United States, Germany or Australia (Chatfield & Reddick, 2017; Ingrams, 2020) rather than in unitary states such as Finland and Sweden, since certain autonomy of local authorities from federal or national agencies could be quite significant (see Fig. 3.4) to distinguish completely different channels of political communication that presumably should exist there, for example, e-federalism and e-centralism (Kassen, 2018). *In this regard, it seems interesting to understand how the concept of*

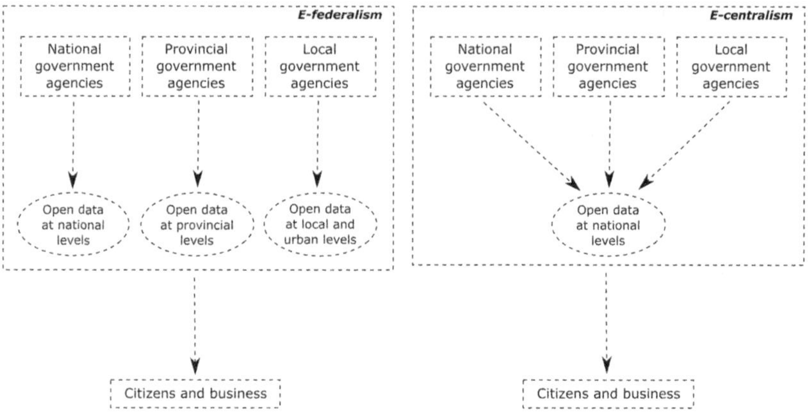

Fig. 3.4 Understanding the implications of different institutional settings in e-federalism and e-centralism. (Source: Own illustration)

open data government is realized in an institutional context of digital centralization based on empirical observation of actual open data-driven projects. This could hypothetically be present in these unitary states, too.

Furthermore, it is really important to understand how the concept is advanced at local levels itself from a cross-cultural perspective, taking into account the fact that the development of open data projects at local municipalities and counties could play a crucial role in the overall promotion of the movement as a presumably cost-effective tool to boost public sector innovations (Jussila et al., 2019; Kukkamäki et al., 2019). In this regard, both Finland and Sweden provide really promising administrative, geographic and demographic contexts to test the assumptions and they could ideally fit the framework of the institutional analysis. In this respect, it is especially important to highlight the significance of open data-driven communication channels at local levels in the light of the rapid diffusion of social media that presumably helps to promote the concept in small municipalities or counties. *In this regard, the differences between bureaucratic structures in the organization of government and certain autonomy of local authorities from the center in some of these unitary countries could hypothetically create a truly interesting diverse environment for the emergence of various independent open data-driven projects, especially in areas related to the promotion of interactive transactions between local authorities and local businesses communities as well as new collaborative channels of communication that could be generated in the area.*

4.2 *The International Levels of Open Data Government Interactions: Understanding What Are the Ultimate Political and Socioeconomic Implications of Global Cooperation in the Area?*

The international level of open data transactions is a relatively new phenomenon in the development of the concept since it appeared only after such global organizations as the United Nations, the World Bank, the World Economic Forum and recently the Open Government Partnership began to cooperate with national governments all over the world in launching various global open data-driven projects in an attempt to develop universal tools of public administration reforms, assessing the socioeconomic readiness of individual nations to adopt e-government and e-participation technologies, comparing their achievements and results in the area at the global level and so on. In this respect, the open data governance agenda could provide a really promising platform for global cooperation of various nations at the international arena (Kassen, 2015). For national governments, participation in such projects as the Open Government Partnership or global e-government forums could be a matter of national prestige, a certain indication of its political commitments to advance digital democracy. In addition, international cooperation in the area implies good opportunities for exchanging knowledge and expertise as well as sharing best practices in the open data area, which, in turn, means better options to choose presumably the most appropriate and previously well-tested models to follow. *In this regard, this certain type of standardization and universalization as a result of the global cooperation between national governments in the area could help to generate generic open data realization tools and, even sometimes, universal technological platforms such as standardized open data portals, blog-platforms, wiki-technologies and so on. It seems interesting to understand how the public values of open data governance are institutionally promoted in various government agencies as certain benchmarks of progress in the area and how the actual proliferation of open data technologies through various cross-border platforms affects their adoption by individual nations.*

The answers to these questions could help to explain in an extrapolated manner the practical interest of many national governments in such international communication platforms and cross-border collaborative initiatives. It is especially important to understand how the cooperation of individual countries with these international projects, which could be tentatively indicated as *a global open data movement*, could provide an additional background for the collaboration of not only national governments as key political actors in the area but also other players such as non-governmental organizations, businesses, mass media and developers, who may also play important socioeconomic roles in this global networking. For example, the cooperation of the Open Knowledge Foundation, a universally prominent non-governmental organization, which promotes open data globally with various national and even local universities, research centers, governments and even individual citizens, or the cooperation of the GovLab, a globally well-known network of open data experts with developers from various national and local open data projects, might be regarded as a striking example of promising international collaborative processes in the area. *In this regard, it seems interesting to know more about the implications of such international backgrounds on the development of open data governance in a cross-country perspective, especially in an attempt to understand the push and pull factors that presumably may affect the desire of national authorities to join the global open data movement.*

5 Conclusion

The stakeholder and institutional analysis of open data governance as a key conceptual and, more importantly, ideological platform to promote both politically and socioeconomically the phenomena of open data has demonstrated that any further exploration in the area would academically be more productive and interesting for, first of all, e-government policymakers and practitioners, if the focus of the following empirical analysis in the research is specifically paid on the investigation of actual collaborative policies that have recently been pushed forward at different institutional levels and lessons that have been learnt by different nations in the area. It is also important to understand the implications of surrounding institutional environments on the development of the open data movement by identifying and analyzing in a comparative manner the key driving forces and challenges various actors face on the way, ideally in different administrative, political and socioeconomic contexts. The analysis of actual *open data-driven projects* that could provide practical evidence about the promising venues of the concept to advance civic engagement, partnership and peer-to-peer networking among various stakeholders both public and private is another interesting area for related case study research.

References

Abu-Shanab, E. A. (2015). Reengineering the open government concept: An empirical support for a proposed model. *Government Information Quarterly, 32*(4), 453–463.

Akcagündüz, E. (2021). Transformation in the delivery of public social services and e-government systems: A study on the Turkish ministry of family, labor, and social services. In *handbook of research on policies, protocols, and practices for social work in the digital world* (pp. 339–354). IGI Global.

Chatfield, A. T., & Reddick, C. G. (2017). A longitudinal cross-sector analysis of open data portal service capability: The case of Australian local governments. *Government Information Quarterly, 34*(2), 231–243.

Feeney, M. K., & Welch, E. W. (2016). Technology–task coupling: Exploring social media use and managerial perceptions of e-government. *The American Review of Public Administration, 46*(2), 162–179.

Frank, M. T., & Zander, S. (2017). The linked data Wiki: Leveraging organizational knowledge bases with linked open data. In *International joint conference on knowledge discovery, knowledge engineering, and knowledge management* (pp. 294–319). Springer.

Gasco, L., Schifanella, R., Aiello, L. M., Quercia, D., Asensio, C., & de Arcas, G. (2020). Social media and open data to quantify the effects of noise on health. *Frontiers in Sustainable Cities, 2*, 41.

Geiger, C. P., & von Lucke, J. (2011, May). Open government data. In Conference for e-democracy and open government (p. 183).

Höchtl, J., & Reichstädter, P. (2011). Linked open data-a means for public sector information management. In *International conference on electronic government and the information systems perspective* (pp. 330–343). Springer.

Ingrams, A. (2020). Organizational design in open government: Two cases from the United Kingdom and the United States. *Public Performance & Management Review, 43*(3), 636–661.

Joseph, B. K. (2019). Blockchain for open data–exploring conceptual underpinnings and practice. In *Governance models for creating public value in open data initiatives* (pp. 161–175). Springer.

Jussila, J., Kukkamäki, J., Mäntyneva, M., & Heinisuo, J. (2019). Open data and open source enabling smart city development: A case study in Häme Region. *Technology Innovation Management Review, 9*(9).

Kandhro, D., Memon, R., & Shaikh, A. (2013). E-government using grid technology: Developing a grid framework for G2G e-communication and collaboration system. In *E-government using grid technology: Developing a grid framework for G2G E-communication and collaboration system* (pp. 8–15). Asian Institute of Advance Research and Studies.

Kassen, M. (2014). Globalization of e-government: Open government as a global agenda; benefits, limitations and ways forward. *Information Development, 30*(1), 51–58.

Kassen, M. (2015). E-government as a universal political value. In *Understanding systems of e-government* (pp. 1–26). Rowman & Littlefield.

Kassen, M. (2016). Key drivers, stakeholders and challenges of the e-government movement in Kazakhstan: Recommendations for practitioners. In *E-government in Kazakhstan* (pp. 101–119). Routledge.

Kassen, M. (2018). Open data and its institutional ecosystems: A comparative cross-jurisdictional analysis of open data platforms. *Canadian Public Administration, 61*(1), 109–129.

Kassen, M. (2019). Promouvoir la coopération publique au sein du gouvernement: principaux déterminants, réglementation et obstacles du mouvement de collaboration électronique au Kazakhstan. *Revue Internationale des Sciences Administratives, 85*(4), 757–777.

Kassen, M. (2020). Open data and its peers: Understanding promising harbingers from Nordic Europe. *Aslib Journal of Information Management, 72*(5), 765–785.

Kubler, S., Robert, J., Neumaier, S., Umbrich, J., & Le Traon, Y. (2018). Comparison of metadata quality in open data portals using the Analytic Hierarchy Process. *Government Information Quarterly, 35*(1), 13–29.

Kukkamäki, J., Jussila, J., & Mäntyneva, M. (2019). Open data enabling networked e-government: Case Tavastia events. In *ISPIM conference proceedings* (pp. 1–12). The International Society for Professional Innovation Management (ISPIM).

Mergel, I., Kleibrink, A., & Sörvik, J. (2018). Open data outcomes: US cities between product and process innovation. *Government Information Quarterly, 35*(4), 622–632.

Obama, B. (2009). Memorandum for the heads of executive departments and agencies. *Presidential Studies Quarterly, 39*(3), 429.

Pereira, R. L., Sousa, P. C., Barata, R., Oliveira, A., & Monsieur, G. (2015). CitySDK Tourism API-building value around open data. *Journal of Internet Services and Applications, 6*(1), 1–13.

Sadiq, S., & Indulska, M. (2017). Open data: Quality over quantity. *International Journal of Information Management, 37*(3), 150–154.

Sebetci, Ö. (2015). A TAM-based model for e-government: A case for Turkey. *International Journal of Electronic Governance, 7*(2), 113–135.

Sebetci, Ö., & Aksu, G. (2014). Evaluating e-government systems in Turkey: The case of the 'e-movable system'. *Information Polity, 19*(3, 4), 225–243.

Yıldırım, S., & Bostancı, S. H. (2021). The efficiency of e-government portal management from a citizen perspective: Evidences from Turkey. *World Journal of Science, Technology and Sustainable Development, 18*, 259–273.

CHAPTER 4

Open Data Governance in Finland: Understanding the Promise of Public-Private Partnerships

Abstract The main purpose of the chapter is to understand the important role of professional collaboration among key stakeholders of the open data movement and analyze a set of typical drivers and challenges associated with the promotion of the concept in both public and private sectors of national economy by resorting to an interesting case of Finland, which has recently adopted various comprehensive open data realization strategies and networking initiatives. These initiatives were aimed at, first of all, outreaching the public values of open data and propagating the political and socioeconomic benefits of its further processing and reuse in various third-party projects and collaborative platforms. The main finding of the research suggests that the active promotion of the concept in private and non-governmental sectors leads to the creation of new market niches where a promising phenomenon of the open data-driven entrepreneurial and philanthropic activity of middlemen emerges, dramatically changing the traditional ways to promote public sector innovations.

Keywords Open data • Finland • Public-private partnerships • e-Democracy • Civic engagement • Innovations

M. Kassen, *Open Data Governance and Its Actors*, Studies in National Governance and Emerging Technologies,
https://doi.org/10.1007/978-3-030-92065-4_4

1 Introduction: Setting the Research Agenda

The main purpose of the research in this chapter is to study a multidimensional and multi-institutional nature of the open data governance concept that could hypothetically provide a promising collaborative platform to bring together the interests of various stakeholders in advancing public sector innovations, namely government agents, citizens, developers, nongovernmental organizations, mass media and businesses alike and understand an important role of the public-private partnership and successful business models in the area. In this connection, this research presents a case study that investigates open data phenomena in the context of a country that is universally well-known today as a global pioneer of the open knowledge movement that explicitly propagates the public value of free access to information as a fundamental pillar of modern democracy.

1.1 Open Data as a Collaborative Political Phenomenon

The key benefit of the open data concept is that it could potentially not only promote the transparency of government through the proactive publication of various public records and datasets but also make it presumably more effective and collaborative due to civic engagement and close networking with the non-governmental and private sectors in launching independent open data-driven projects. In this regard, the author of the research tries to understand in this chapter what kind of political and socioeconomic drivers and challenges Finland faces today in the diffusion of the open data concept among various stakeholders, namely government officials, developers, businesses, journalists, academia and nongovernmental organizations, and how it overcomes them, especially in areas that are related to the publication and, more importantly, to the reuse of datasets by third parties in various collaborative projects and independent civic platforms. The main purpose of the research is to study a multidimensional and multi-institutional nature of the open data concept that could hypothetically provide a promising collaborative platform to pull together the interests of various public and private stakeholders in advancing public sector innovations by resorting to an interesting case of the nation that has recently adopted various comprehensive realization strategies and networking initiatives in the area. The case of Finland seems especially appealing, taking into account the recent rapid promotion of this small Nordic country in various global open data ratings, according to various international rankings (Global Open Data Index, 2019; OECD, 2019; UN, 2020). Moreover, in 2020, Finland was ranked fourth among all nations in the prestigious Global Open Data Index (WJP OGI, 2020).

1.2 Understanding the Research Perspective

This research presents a case study that investigates a promising collaborative potential of the open data concept from stakeholder and institutional perspectives. Presented in more a holistic manner, especially in analyzing new open data-driven professional platforms and related communication activities that bring together the interests of various players as diverse as government agencies and journalistic communities, developers and academia, charities and businesses, the chapter seeks to better understand the fundamental factors of open data governance as a collaborative political phenomenon. In this regard, the analysis is presented both as a policy review and as an agenda-setting research, which is intended for a global community of open data experts and all those who are interested in studying the phenomenon from the perspective of its development in open societies. Therefore, in the conclusion parts of the research, a number of practical recommendations are provided for both policymakers and practitioners. The author also tries to engage its readers into the topic with a lot of unique visualizations and illustrations that could help them to better understand the multidimensional nature of open data phenomena in this Nordic country. Moreover, considering that the chapter also seeks to propose a universal framework for the case study research that could be used in investigating the phenomena not only in Finland but also in the context of other post-industrial societies, especially in analyzing the political and socioeconomic drivers and barriers in the adoption of related collaborative and civic engagement platforms and startups, the results of the research could also be used in evaluating and improving the operation of various open data-driven projects all over the world, providing an invaluable and unique stakeholder's perspective on the development of this promising concept.

1.3 Finland as a Poster Child of Technological Modernization

Finland is globally regarded as one of the most successful examples of the nation that during last three decades has managed to implement various technology-driven projects in many areas of national economy whether it is in building innovative industries, promoting cutting-edge research and education policies in universities and laboratories, advancing various public sector reforms or propagating extremely cost-effective e-administration systems. This Nordic country is also universally well-known for its developed infocommunication sphere and a number of global technological corporations, which have recently demonstrated an unprecedented growth in terms of both increasing expertise and boosting technological progress in a really wide range of areas such as software industry, telecommunications, electronics, nanotechnology, medicine and so on. This truly enriching and invigorating atmosphere of innovation and knowledge nurtured by various players both in public and in private sectors of economy has benefited the e-government field, too, presumably making the public sector reforms more effective due to the increased focus on innovative solutions and, more importantly, professional networking platforms, including in the open data area, where this nation has demonstrated really impressive results not only in publishing proactively a plethora of open datasets but also in applying various interesting strategies and tactics on how to better promote the concept among its key target audiences, that is, in public, non-governmental, private and technological sectors of economy, for example, among public agencies, developers, businesses and citizens alike, especially in encouraging them to generate, reuse, process further and even publish their own open data in special public repositories (see Fig. 4.1). In this respect, the author of the book argues that the professional networking among the key stakeholders of the open data movement has played a crucial role in the unprecedented diffusion of the idea in Finland as a new highly effective collaborative platform to advance public sector reforms and innovations.

1.4 Understanding the Academic Background

Even though there are many case studies on the development of the open data concept in various countries, including in Finland, most of the research has been concentrated on the investigation of the phenomenon from the perspective of benefits and challenges that it may bring to the public sector, rarely focusing on the analysis of new promising communication processes that emerge between the key stakeholders of the open data government movement itself, for example, in studying the implications of public-private partnerships in only particular areas of open data-driven collaboration such as among public agents themselves (Hyvönen,

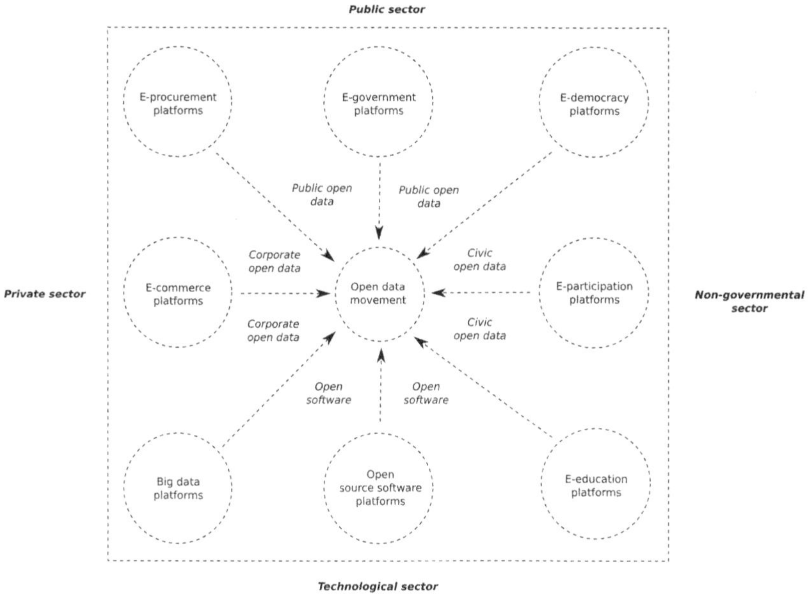

Fig. 4.1 The multidimensional phenomena of the open data generation movement. *Source: own illustration*

2020; Lehtonen, 2011), between business and academia (Lehtomäki et al., 2015; Perkmann & Schildt, 2015), ICT developers and public agents (Hielkema & Hongisto, 2013; Hyvönen, 2020; Kassen, 2018a), citizens and non-governmental organizations (Shiramatsu et al., 2015; Sihvonen & Sipilä, 2014; Still et al., 2013), mass media (Aitamurto, Sirkkunen and Lehtonen, 2011; Kinnari, 2013), open data entrepreneurs and innovators (Magalhaes, Roseira & Strover, 2013; Lindman, Kinnari & Rossi, 2014; Kassen, 2018b; Ylipulli & Luusua, 2020; Pynnönen et al., 2021) and so on. In this regard, this research presents a case study that investigates the promising potential of open data-driven professional networking platforms and related communication activities in a more holistic manner in an attempt to better understand the fundamental factors of the open data movement as a collaborative political phenomenon.

2 The Methodology of the Research

The rapid widespread diffusion of open data phenomena all over the world and, more importantly, an interesting trend in recent emergence of new professional communities and promising networking activities among various stakeholders, both public and private ones, around the open data movement are ripe for close investigation and policy analysis through the prism of actual cases in a real political practice. This was partly one of the reasons why the author of the book decided to resort to case study research, applying in a successive order the combination of three methods of research, namely the context, policy and content analysis. Consequently, this phenomenon is explored in Finland through the prism of key stakeholders, drivers and challenges of the open data movement with the main goal of the whole investigation is to understand how this promising idea is developing in an unusual context of an advanced digital society that is globally well-known for its excellent sophisticated solutions in open data areas, especially demonstrating really impressive results in promoting open data-driven platforms, according to various international rankings (Global Open Data Index, 2019; OECD, 2019; UN, 2020). In this regard, the context analysis of country backgrounds is helpful in studying the political and socioeconomic environments that are hypothetically conducive to the development of the concept and assessing the readiness of key stakeholders to adopt and lead the open data movement. The next research method, a policy analysis, is based on the close investigation of the realization and regulation strategies, adopted by the Finnish government mostly at the central level, as well as the observation of interesting patterns in political communication processes among the key policymakers of the open data movement, which could help to shed light on the role of private and non-governmental sectors in the diffusion of the concept in this Nordic country and reveal related networking activities among various stakeholders, especially developers, journalists, non-governmental organizations, academia, businesses and technically savvy citizens. Finally, the content analysis of open data-driven projects launched independently by the members of civil society could help to better understand the civic component of the open data paradigm. In general, the combination of these three methods will be useful in understanding better political, social and economic implications of the country's context on the development of the open data movement in Finland and locating the most important institutional drivers and challenges that the policymakers face today in the realization of related digital collaborative and participatory platforms.

3 The Country Context Analysis: Understanding Political and Socioeconomic Implications of Open Data-Driven Public-Private Partnerships

3.1 The Economic Environment: Focusing on Mutually Beneficial Public Sector Innovations

A really impressive economic development of Finland during the last three decades unprecedented in terms of both the scale of the reforms and the engagement of public sector into the realization of various industrial projects, especially in areas such as shipbuilding, electronics and software production, has created a really diverse and sustainable market, which provides a truly favorable environment to promote the ICT-driven development agendas almost in all niches of national economy, including in furthering public sector reforms. Another aspect of the favorable economic background that has accompanied the public reforms is a knowledge-based education policy and intensive research activity in various Finnish institutions and universities where the primary focus on fundamental studies and empirical exploration is deeply embedded into the education system. The strategic reliance of many local telecommunication companies on the output from the research and development activity (R&D) implies also a focus on innovations and technological advance in a wide range of electronic industries, which results in the creation of a really favorable political environment for the promotion of various startups and initiatives in the area, including in the open data area, which now is increasingly being regarded in Finland as a completely new business niche (Jaakkola et al., 2014). The concept itself is regarded by representatives of both government agencies and private corporations as an additional way to save public funds and boost dramatically domestic economy due to open data-driven public sector innovations and new public administration since the reuse of government datasets in the private sector also implies an active generation of various third-party public startups and business solutions (see Fig. 4.2). In this regard, the traditional cooperation between educational

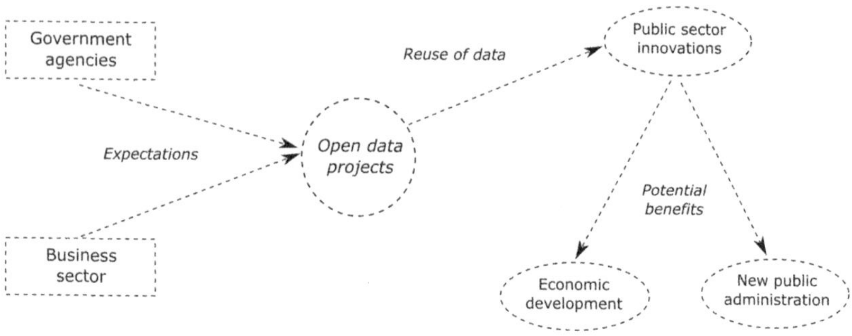

Fig. 4.2 Potential economic benefits of open data-driven public-private partnerships. *Source: own illustration*

institutions and local technological companies, especially in launching various training programs and research activities, creates an additional impetus to advance socioeconomic inquiries in this direction, too.

3.2 The Social Environment: Harnessing the Promise of Civic Engagement and Collaboration with Private Sector

Finland is globally well-known as a socially oriented state that has succeeded in promoting and developing further the Scandinavian model of the social welfare system during the last three decades (Timonen, 2003) that has a huge indirect multiplicative impact on the development of many spheres, including public administration and governance, benefiting the public sector in terms of cost-effectiveness and collaboration due to the increased level of readiness of all stakeholders to comprehend new ideas and capability of well-educated and generally more prosperous citizenry to adopt and use various participatory public and private platforms, regardless of their levels of income and social status. In this respect, Finland could be regarded as a classical example of the country where the high standards of living due to the socially oriented economic policy of the state (Kettunen, 2001) and explicit focus on matters of social justice result in the creation of better opportunities to advance open data and other related technological projects, indicating to an assumption that more even distribution of wealth and income leads to better achievements in the area and is closely related to the socioeconomic readiness of local population to harness fully the potential of the concept in all its diversity whether it is in promoting participative e-government or its last interpretation open data, increasing collaboration between civic society, businesses and government agents or reusing open data for third-party initiatives. In the end, more educated, wealthy and well-informed citizens could better contribute to the ultimate goal of the open data, that is, on how to engage people in building their own information products from public government datasets, which would presumably benefit both citizens and government, making the latter more efficient and less costly.

3.3 The Political Environment: Advancing the Open Data-Driven Political Agenda among Key Policymakers

The political environment is probably one of the most important factors in the development of the open data concept in Finland. The readiness of the Finnish government to promote and support various open data startups, especially at the national level (ODP, 2020), results in a really favorable political atmosphere for all stakeholders to launch such projects. Another aspect of the policy is a strong tradition of representative democracy and political pluralism embedded into the political system (Bengtsson & Mattila, 2009), which has been actively promoted and supported by all political forces since the moment the nation gained its independence in 1917. The traditions that value openness and transparency in government proclaim explicitly the right of citizens to be engaged in associations and access the official datasets and information, which creates a really robust political basis of civic participation that is truly conducive for the development of various autonomous political organizations and civic engagement projects whether it is in creating really diverse and polarized factions in the national parliament represented by a wide range of political forces and ideologies, operating autonomous, self-sufficient and sustainable trade unions or promoting various non-governmental organizations and e-participation projects, including in the open data sphere. In this respect, representatives of both public and private sectors clearly understand the political and economic value of the concept (Lehtonen, 2011), which realization promises new opportunities to boost national economy and strengthen civil society through the open data-driven innovations in the public sector, in general, and the increased level of transparency and civic engagement in the e-government area, in particular.

3.4 The International Environment: Pioneering the Global Open Data Movement

The international dimension arguably plays one of the most important roles in the diffusion of the open data governance concept at the planetary scale (Kassen, 2019). In this regard, there are two major factors that could be regarded as drivers of the open data movement in Finland: the first one is the commitments that the Finnish government had to take to promote open data and transparency in government when it joined in 2013 such a global intergovernmental organization as the Open Government Partnership (OGP, 2013). The second one is the international image of the country as one of the well-known leaders in technology-driven government solutions, which provides additional incentives to advance open data as a new innovative platform to further various open data projects in cooperation with other stakeholders such as citizens, academia and businesses, truly demonstrating a multidimensional phenomenon of the movement where both internal and external factors of political communication play an equally important role of motivation to advance open data-driven projects and related civic engagement platforms. For example, any executive decisions made by national authorities in the area are counterbalanced by feedback, critics and ideas that come from civil society, leading to a more collaborative generation of actions, plans and commitments that the country has to take as a full member of this global partnership. Moreover, the open data movement itself is promoted due to close public-private partnership, civic engagement and cooperation in the area since the government publishes datasets and representatives of civil society harness or reuse them to create new open data-driven products and services. In this regard, the whole process of such collaboration between these stakeholders and, more importantly, the production of new civic data-driven initiatives in the area are monitored by the international community of e-government experts in this intergovernmental organization.

It is interesting to note that even when national authorities decided to join the global initiative in order to ensure transparency of the procedures, they resorted to a promising potential of collaboration and civic engagement in drafting the commitments that the country had to take by organizing special workshops with various stakeholders of the open data movement in order to receive feedback from local residents, data scientists, e-government professionals, policy experts and non-governmental organizations, including by using e-participation and other technology-driven platforms and citizen-sourcing projects. Not all, even the most established democracies have resorted to the same collaborative practice as Finland in drafting the country's action plans in the open data sphere and discussing political commitments toward the Open Government Partnership in such close cooperation with all members of civil society. This was one of the reasons why Finland is universally regarded as one of the global pioneers in promoting the political ideology of open data and as a country it occupied a prestigious fourth place in the Open Government Index in 2020. This global index is calculated by experts from another independent player in the area—the World Justice Project, which is a well-known international non-governmental organization that propagates the values of public justice and rule of law on a planetary scale (WJP OGI, 2020). In this regard, the strong commitments of the nation to promote the transparency of government through the use of information and communication technologies, including in developing various government-related projects, create an additional favorable environment to advance the open data concept, too, especially at the national level. Taking into account the fact that Finland is a typical unitary state and many initiatives in the area are of the traditional top-down nature, it should presumably make the realization of open data-driven projects smoother and easier from both bureaucratic and organizational perspectives.

4 Open Data Regulation: Adopting the Oldest Traditions of Freedom of Information to the Needs of the Open Data Movement

The Freedom of Information Act (FOIA) is globally regarded as one of the most important legal documents that regulate the public access to government files and information and could be applied to the open data concept, too, taking into account the similarity of regulation in the area. In this regard, it is necessary to note that the freedom of information law was enacted in Finland as early as 1766 when it was in the political union with the Kingdom of Sweden. The document that was officially named as the Act on the Freedom of Publishing and the Right of Access to Official Documents (De Smaele, 2007) is regarded as the oldest FOIA in the world (Ackerman & Sandoval-Ballesteros, 2006). In its modern sense, the new law that was adopted in 1951 as the Act on the Publicity of Official Documents (Berliner, 2010; Saario & Ginsburg, 1965) proclaimed a universal right of citizens to access various government documents and records regardless of the category of information whether it is provided at national, regional or local levels of governance. This document had provided a really strong instrument of civic control and scrutiny in the public sector reforms long before the emergence of open data as a political philosophy in 2009 by granting access to government records of various kinds to citizens, non-governmental players, mass media and researchers from local academic institutions and think tanks, reinforcing later the fundamentals of the open data government concept and digital democracy in this country. Furthermore, the amendments adopted in 1999 resulted in the revision of the law now being called as the Act on the Openness of Government Activities (Kuutti et al., 2012), widening the regulatory instruments not only in relationship to government agencies but also to the corporate world since a wide range of public datasets were also generated by businesses, and therefore, should have been regulated, too. The revised FOIA provided new incentives to develop a multidimensional tool of public access to government information and even public corporate datasets that are somehow related to the aggregation of information on any citizen, truly putting Finland on the forefront of the freedom of information regulation policymaking in the world.

5 The Policy Analysis of the Open Data Governance Movement in Finland: Key Stakeholders

5.1 *Understanding the Role of Government Agencies*

The role of the public sector in the promotion of open data policies in Finland is truly crucial since it is a unitary state and all administrative processes at all institutional levels are highly centralized there. The central platform of open data (*avoindata.fi*), which currently hosts more than 1800 various datasets from more than 700 public and semi-public stakeholders, is directly supported by the government. There are many public agencies that are actively involved in disseminating the concept at both national and local levels, such as the Finnish Digital and Population Data Services Agency (FDPDS, 2020), Ministry of Transport and Communications (MTC, 2020) and Ministry of Justice (MJF, 2020), but the overall political and administrative oversight and operation of the official open data project is carried out by the Ministry of Finance (MFF,

2020). The national statistical office is another public organization that publishes a wide range of open datasets and statistical information on various topics and strategic areas of the national economy (Statistics Finland, 2020). The National Land Office of Finland, which provides a variety of topographic information and maps (NLOF, 2020), is also an active player in the area. Another important player in the open data movement in the public sector is the Finnish Meteorological Institute, which provides datasets for free to businesses and the civil sector (FMI, 2020), leading to the creation of a second market for various meteorological applications both in Finland and abroad. All of these central and local government agencies and civil society organizations provide a conducive political and administrative environment to promote the open data movement as a truly collaborative project aimed primarily at reusing government datasets in various third-party projects, even though they create separate depositories at national and local levels.

The quasi public sector is also playing an important role in the advancement of open data in various sectors of economy. For example, the government-related Forum Virium Helsinki is another public platform that actively promotes the concept at local levels in the Helsinki metropolitan area in various local public and academic institutions in close collaboration with the city municipality, universities, businesses and other stakeholders (Forum Virium, 2020). Founded in 2006 and bringing together representatives from dozens of public and private sectors of the Finnish economy, such as local governments, telecommunications companies, local electronic and print media, development foundations and charitable organizations, on its governing council, it provides a wide range of services, consultations and statistics to open data actors and other innovators in the Helsinki metropolitan area. It also helps to organize various seminars and webinars on innovative topics, especially to promote such emerging promising concepts as smart cities, e-government, e-participation, e-commerce, e-transportation, e-procurement, e-healthcare, technology-based local economy, peer-to-peer governance, new interactive media, real-time economy and so on (see Fig. 4.3).

One of the most interesting projects, currently being carried out by the Forum Virium organization in partnership with other regional development funds such as the Tampere Regional Economic Development Agency (Tredea, 2020), is an initiative called the Six Cities Strategy, which aims to stimulate an innovative economy and new developments in the state sector by harnessing the promising potential of open data as a cost-effective

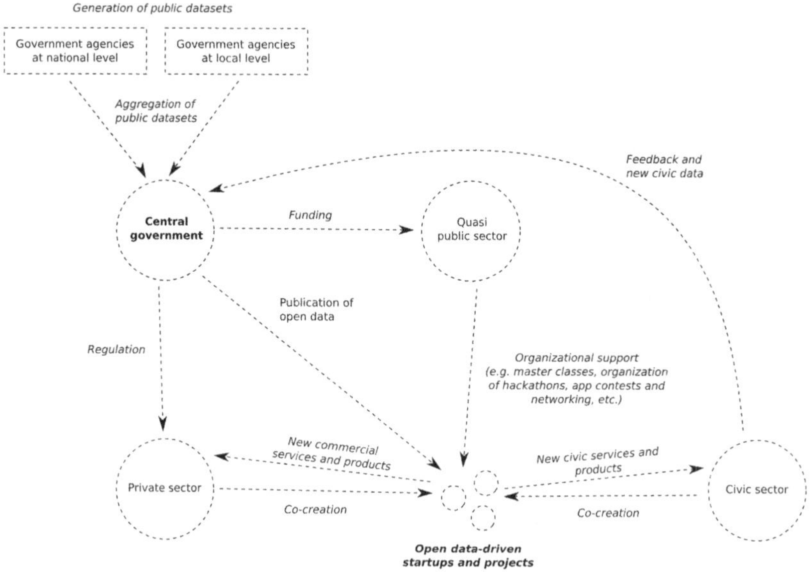

Fig. 4.3 The focal role of the public sector in initiating and promoting the open data movement. *Source: own illustration*

element of urban policy and public sector innovations in cities such as Helsinki, Turku, Tampere, Espoo, Oulu and Vantaa (6Aika project, 2020), which could presumably help Finland to become more competitive at the international arena, since the lion's share of national income is generated in these six largest urban regions of the country. In this regard, Forum Virium is not only a promoter and supporter of various digital open innovations in economy and government but it is also a collaborative and networking venue for various stakeholders in the areas to cooperate and exchange ideas in a friendly and transparent manner.

5.2 *Understanding the Role of Business Communities*

The business sector is an active promoter of various open data projects and platforms in Finland. The organization of various collaborative and networking events in the area is often directly supported by a wide range of local telecommunication companies and software startups. For example, a somewhat win-win strategy adopted in Finland, which is aimed at benefiting all stakeholders through the development of open data-driven software algorithms as a business model in commercializing the related government digital services, has even resulted in the creation of a specialized market in the area. In this regard, such a company as the Cloud'N'Sci Ltd. that designed a unique technological concept, which they called as the Algorithms-as-a-Service (AaaS), provides a very interesting precedence to promote the concept of open data in the corporate world, potentially revolutionizing the idea of open data by offering a working mechanism not only on how to reuse open data and launch related projects in furthering the public sector innovations but also on how to economically benefit from its implementation, providing additional incentives for such important stakeholders as developers and businesses in adopting and utilizing originally proprietary and, therefore, presumably more cost-effective mechanisms to develop commercial open data-driven products and digital services. The developers of this revolutionary platform claim that the predeveloped algorithms of open data-driven projects as a separate product could be interesting for both businesses and developers due to the lower costs for research and development, better quality in management of datasets and other interesting features (The Cloud'N'Sci.fi project, 2020). Another Finnish company that participates in the promotion of the concept and develops other related commercial networking projects in the area is Agience Oy Ltd., which was founded in 2013. It has already initiated such business startups in the area as Agile Science and Innovation Compass projects with an ultimate goal to network academic researchers and corporations on a commercial basis (Agience, 2020), recognizing a huge economic potential of the open data concept to boost innovations and collaborative processes in the market.

5.3 Understanding the Role of Independent Software Developers

The networking among developers is really crucial since the promising potential of the open data concept could be realized only if it is processed and used by ICT professionals in their third-party applications and software programs. In this regard, the core of the successful realization for many private and public open data-driven projects and platforms in Finland is a really diverse and wide network of local developers whose expertise and readiness to share the best solutions in the area is difficult to overestimate. In this regard, it is necessary to note that the popularity of open-source software platforms and altruism as a philosophy among many technically savvy citizens who often call themselves independent developers has contributed to a truly enriching atmosphere of collaboration and participation surrounding the open data concept in this Nordic country. Such networking platforms for ICT developers as, for example, government-affiliated Valtori center (Valtori, 2020) or Helsinki Loves Developers platform (HLD, 2020) provide a wide range of consultative, technological and crowd-sourcing services to national and local authorities in reforming public administration, including in launching and maintaining various open data-driven projects and startups.

In this regard, the close cooperation of many independent developers with local and global non-governmental organizations helps to negate the operation costs in creating and maintaining technologically the open data-driven projects due to new opportunities of co-funding and, more importantly, access to innovative networking venues, which, in turn, allow to use universal highly standardized developing platforms and open-source software, the creation of which is supported and financed by global non-governmental players, including in managing the massive stockpiles of open datasets. In this respect, such universal data management systems as, for example, the Comprehensive Knowledge Archive Network (CKAN), which has been developed by the Open Knowledge Foundation (CKAN, 2020), could be regarded as one of the most cost-effective attempts to provide a common standardized technological and networking platform to promote the open data concept all over the world. Another aspect of such close cooperation between developers and non-governmental players is the organization of special events and networking platforms aimed at specifically boosting innovations in the open data sphere. For example, various contests and hackathons organized in Finland such as the Open Finland Challenge (OFC, 2020), the traffic app competition and 48-hour Ultrahack hackathons (Ultrahack, 2020) help not only to find the best technological solutions in promoting the concept but also engage young developers to the open data domain. These app contests are the real incubators of new ideas and innovations in the open data area that generate a plethora of open data-driven solutions and projects, advancing Finland in the development of the concept as a collaborative and entertaining platform for professionals and young developers, students and researchers, journalists and government practitioners, eventually benefiting every stakeholder of the open data movement since the promotion of peer-to-peer networking solutions is crucial for the diffusion of independent civic engagement projects in the area. In this respect, one of the most interesting ways to promote the concept was offered by the organizers of the Datademo project, which claims to be a participatory budgeting platform that allows the contestants themselves to choose the winner of the contests (Datademo, 2020), revolutionizing itself the concept of open data apps competition as a collaborative networking process rather than just a rivalry.

Various networking events such as professional meetings, webinars, hackathons, brainstorming seminars and conferences help to propagate the open data concept among various stakeholders in Finland especially at the national level by inviting government agents, developers, non-governmental organizations, journalists, academia and students as well as technically savvy citizens to use open data in different instances and for various purposes whether it is in building third-party applications based on utilizing the citizen-sourcing concept, participating in gathering additional information and collecting new data for the existing open data-driven platforms and projects, informing the public about the successful stories in open data areas through mass media and social networks or simply studying, doing research in the area and sharing its results with other peers. The promotion of such networking platforms as, for example, the Mindtrek or Apps4Finland projects, which aim not only at rewarding the best open data-driven projects created and launched by local developers but also at attracting their attention to the latest events in the area, really helps to unite various groups of developers and ICT professionals such as software programmers, designers, testers, system administrators, engineers and so on (Apps4Finland, 2016; Mindtrek, 2020), leading to the creation of a more collaborative network of local developers and, eventually, driving innovations in the open data area. It is this promising networking phenomenon that promotes the open data movement as one of the most advanced platforms in terms of finding a successful formula of the public-private partnership between government and citizens where developers play a crucial role of the technological bridge, the fact that is often ignored or underestimated in the e-government literature, which usually pays more attention to the related communication processes between government agencies, businesses and citizens as presumably the most important stakeholders of the open data movement.

5.4 *Understanding the Role of the Civic and Non-governmental Sector*

Finland is universally regarded as one of the most open and liberal societies in the world. In this regard, civic and non-government sector are among the most important stakeholders in the diffusion of the open data movement in this country, taking into account the really important role that it plays in promoting the ideas of freedom of information, democracy, political participation and civic engagement. Such non-governmental and non-profit organizations as, for example, the Open Knowledge Finland (OKF, 2020) and the Finnish Association for Online Democracy actively propagate the concept by providing a wide range of consultative, technological, expert and administrative services to other non-governmental organizations, civic developers, government agencies and groups of civic activists as well as offer various funding options for many independent open data-driven projects at both national and local levels.

The Innovation Fund—Sitra—is another important non-governmental player that actively participates in diffusing the open data concept and promoting the sustainable development of Finland (Sitra, 2020), including through harnessing the promising potential of open data to boost local economy, technological innovations and digital democracy. The development of information society, digital government and nurturing a sustainable economic environment for business communities is an ultimate goal of various projects that are currently supported by the fund such as the Protomo project that helps local entrepreneurs with consultations and guides on how to innovate, promote themselves domestically and globally and find new markets (Protomo, 2018) or the Ominvoimin initiative that aggregates and publishes various innovative ideas proposed by citizens themselves (Ominvoimin, 2015), which, along with other collaborative platforms, could be regarded as one of the most effective peer-to-peer innovations, generating community networking projects and attracting various stakeholders from both public and private sectors of the national economy in Finland.

Among other interesting local non-profit organizations, it is also important to note the Avoin Ministeriö, which was launched and based in Helsinki. The platform propagates the ideals of open data governance at various levels by promoting crowd-sourcing, civic engagement and participatory platforms such as open legislation, e-democracy, e-participation, e-collaboration and civic engagement. One of the most famous projects of this non-governmental organization is the cognominal platform—Avoin Ministeriö, which is translated as The Open Ministry in Finnish. The project provides people with an opportunity to propose and promote their own open data-driven initiatives in these areas (Avoin Ministerio, 2020) and also allows visitors to vote for or against various legislative proposals, demonstrating a promising potential of civic engagement, digital democracy and open data-driven participation in public and private areas.

5.5 *Understanding the Role of Academic Communities*

Various universities and colleges in Finland participate in the promotion of various open data projects across the nation. For example, such centers of tertiary education and cutting-edge research in Finland as University of Aalto, University of Turku, University of Helsinki, Tampere, Oulu, Jyväskylä and many other universities and colleges do research in the area and cooperate with both public and private sectors in developing various innovative open data-driven platforms, e-government, e-democracy, e-participation, e-procurement and e-commerce projects by sharing knowledge and providing expertise. A number of technoparks and technological clusters are promoted by local academic communities, which help to advance data-driven innovations among local businesses, especially in attracting the interest of various young startupers and innovators. Universities are also actively generating a wide range of open datasets, related to science and technology, education and research in a wide range of areas and industries, which are later published or shared with public, private and non-governmental sectors of the economy for further reuse by local developers, data scientists and data journalists. The distinctive open knowledge movement in Finland is promoted through open data-driven generation of new ideas for such areas as open science, open software and open innovations (see Fig. 4.4).

5.6 *Understanding the Role of Journalistic Communities*

Data journalists play an important role as intermediaries who often use open and big data for their own journalistic investigations and actively promote the idea of open data governance in Finland, too. Local media communities provide really important information coverage of various events related to the open data movement. In this regard, various local public media corporations often publish their own content as open data in

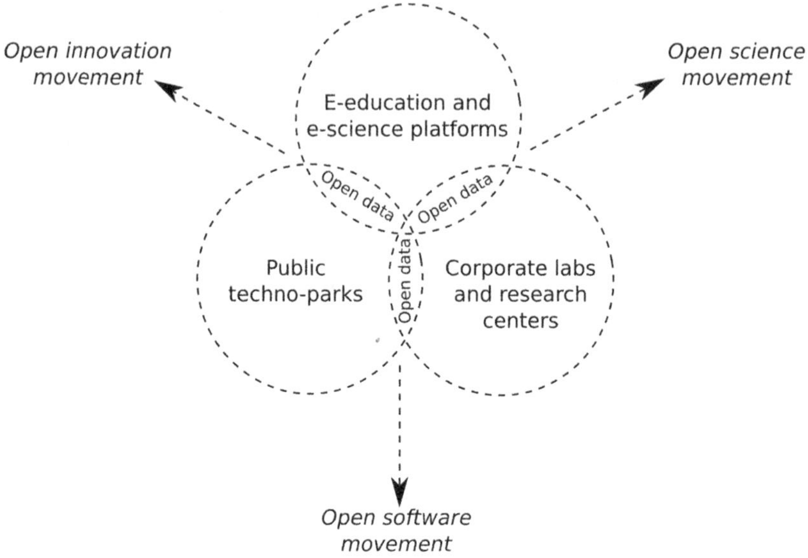

Fig. 4.4 The key elements of the open knowledge movement in Finland. *Source: own illustration*

order to promote its further reuse in third-party applications. The content and datasets could be publicly accessible for developers and technically savvy citizens in an attempt to promote various media-related open data-driven programs and startups in the area. Such topics as parliamentary elections, foreign relations and peacekeeping operations, public budgeting, public auditing, public procurement and so on are among the most interesting areas for local data journalists and innovators to concentrate and innovate in Finland. Various visualization and mapping tools as well as statistical analysis solutions and other media data-driven journalistic products and initiatives are among the most popular that drive the movement in the media sphere. Such Finnish media organizations and conglomerates as, for example, the national largest newspaper—*Helsingin Sanomat*, the national television agency—Finnish Broadcasting Company YLE and one of the largest social media websites in Finland—Suomi24, among others start paying attention to the emerging market for data journalism and even tentatively beginning to publish their own datasets.

6 The Evaluation of Open Data Realization Policies in Finland: Promises and Perils of Semi e-Federalism

6.1 *Projects at the National Level: A Paradoxical Symbiosis of National and Regional Open Data Platforms*

Despite the fact that it is officially a unitary state where much of the political power is concentrated in the hands of central government, which is usually conducive for the development of e-centralism (Kassen, 2015), that is, a single e-government realization policy, Finland has several official open data initiatives at both national and local levels of governance, clearly demonstrating a classical example of semi e-federalism that could be characterized by a presence of multiple regional semi-autonomous ICT-driven public sector projects with a primary dominating role of a single national platform. In this regard, as of 2020 the main official national open data project hosts more than 1800 datasets (Avoindata, 2021), aggregated from various public and private sources, for example, government agencies, universities, libraries, museums, meteorological stations, national companies and private entities, since even individual citizens have the right to not only use open data in creating various third-party applications and related computer programs but also register and publish their own datasets, dramatically widening the traditional understanding of the concept as an only public dimension where government bodies could be regarded as the exclusive legitimate publisher of open data and, therefore, demonopolizing its often special right to publish public information and providing the private stakeholders such as corporations, media and citizenry with an opportunity to participate in the process, too.

6.2 *Projects at Local Levels: Engaging Regions and Municipalities to the Open Data Movement*

Among the largest local open data projects, it is necessary to mention the Helsinki Region Infoshare Project (HRI, 2020), the open data initiatives of the city of Tampere (Tampere Avoin Data, 2020), Oulu (Oulu Avoin Data, 2020) and Jyväskylä (Jyväskylä Avoin Data, 2020). One of the interesting aspects of the Helsinki Region Infoshare project, which serves a huge metropolitan region of the capital city and nearby towns of Espoo, Vantaa and Kauniainen, hosting various datasets from a wide range of sources both public and private ones is that its visitors not only could retrieve and publish their own datasets but also submit data requests in an attempt to draw attention of local government agencies or businesses to a particular area of public administration or economy. Taking into account the fact that much of the information in its portal is also provided in English and it, therefore, could be used also by foreigners in their own data-driven solutions, this innovative project became popular not only among local residents in Helsinki but also foreigners. All these features of semi e-federalism create a unique institutional landscape for the promotion of open data governance at different administrative levels in Finland, namely at the central (e.g. represented by the national open data platform) and local levels (e.g. represented by various open data platforms in such urban areas and cities as Helsinki, Oulu, Turku, Tampere, etc.).

7 The Content Analysis of the Independent Open Data-Driven Projects in Finland

7.1 The rOpenGov Project: Open Data-Driven Collaborating and Networking Initiative

The rOpenGov initiative is a community project that has been actively promoted since 2010 when it was created first with an ultimate goal to promote open collaboration, co-development and networking among data enthusiasts and independent e-government, e-participation, e-science, e-mapping, e-democracy, e-education, e-legislation, e-healthcare, e-procurement and e-commerce software developers. The word R stands for R packages, which are based on R and partly also on C and FORTRAN programming languages. The R language is widely used among proponents of open-source software, especially for statistical computing purposes, socioeconometrics, data visualization, and as such is also quite popular among open data and e-government developers, data scientists and analysts, data journalists, post-graduate students, statisticians and data miners for a wide range of statistical software products and creation of various mobile applications and also could be used for data analysis, academic survey research and data-driven analysis in the review of scholarly information by academic and librarian communities. Being promoted by local developers, many of whom are independent actors who directly contribute to the exchange of ideas and networking between key stakeholders of the open data movement in Finland (e.g. in specialized blogs and interconnected channels on the GitHub professional social networking platform), this promising project is widely used as a venue for data-driven collaboration in many areas and industries in this country and beyond (The rOpenGov initiative, 2020).

7.2 *The Puoluekentta Project: Identifying the Latent Networking Activities among Local Politicians*

The Puoluekentta project that was built on the reuse of election data could be regarded as one of the most successful models in the realization of the open data concept in the political domain since it provides very interesting information on political orientations of the members of the Finnish Parliament as a visualization tool that could be easily comprehensible and usable by the constituents and, more importantly, local media communities, as an analytical tool (Lehtonen, 2011). Using this open data-driven service that was launched in 2011, citizens and, more importantly, journalists and social activists not only could locate the political ideologies of their favorite candidates and public figures but also compare them with other politicians. By filtering related information in accordance with the constituencies, partisanship, political weight, popularity and other factors (The puoluekentta project, 2020), citizens, researchers, media and representatives of local business communities could better understand the political structure of the government they elected. This project contributes to the openness of government and transparency of political processes in this country by informing civil society about the political spectrum, level of polarization and potential groups of interests in current government since Finland is a parliamentary democracy, whose prime minister and its cabinet are directly appointed by the legislative branch of power and accountable before it.

7.3 *The Parliament Explorer Project: Tracking the Political Decision-Making at the National Parliament*

Another open data-driven project aimed at increasing transparency of government is a Parliament Explorer initiative. The app designed for the mobile operating systems is offering its content in both Finnish and English languages, helps citizens to track in an interactive mode the decision-making and political activity of the Finnish Parliament

(Eduskunta) by looking at the government bills that are currently under consideration in the legislature, retrieve information on the partisanship of its key members, know the internal structure of parliamentary committees and lobbying groups, follow the plenary sessions and meetings of deputies and understand the level of political polarization and ideological division of the parliament into party factions, group of interests, unofficial regional representation and so on (The Parliamentary Explorer project, 2020). In addition to open data that is retrieved from the public datasets repositories from the Finnish Parliament (eduskunta.fi), the project also uses data from such sources as the national government (valtioneuvosto.fi), the presidential office (presidentti.fi) and Ministry of Justice (finlex.fi). One of the interesting aspects of the startup is that it not only uses open data from various national datasets and portals, processes and provides it in a easily comprehensible form but also retrieves various types of information from official social media accounts of the members of the parliament, harnessing a promising democratic potential of two, indirectly related to the open data concept, ideas of civic engagement and e-participation in a single tracking and networking platform (Forum Virium, 2015). Another promising aspect of the open data-driven application is that it also publishes contact details of key members of parliament and allows tracking their comments in various social networking platforms, which presumably could help to increase the level of political communication with journalistic communities, peers and, more importantly, constituents.

7.4 The Päätökset Project: Boosting Political Communication with Citizens and Local Governments

One more networking project that is aimed at boosting the level of political communication and digital participation in Finland is called the Päätökset initiative. One of the primary goals of the project is to increase dramatically the transparency of the local government in the capital city of Helsinki by informing local residents about all political and administrative decision-making processes in the city hall (Päätökset, 2020) and, more importantly, on what kind of basis and reasons they are made. It allows filtering the decisions in accordance with government agencies that participated in their adoption and even provides access to the legal acts that officially document them in a retrospective manner. Another interesting feature that helps citizens to easily track the government activity in the area is a sophisticated visualization and mapping tool embedded into the system, which allows not only to see the decisions made in a particular neighborhood of the city but also provides a strong instrument of political analysis for local journalists, researchers, non-governmental organizations and other interested stakeholders. The platform also allows to receive tweets and updates on government-related decision-making procedures that are currently under consideration in local municipalities in an interactive mode, which could help the local constituents to track their activities and react accordingly by commenting and posting in social media networks and platforms, dramatically increasing the intensiveness of political communication between regional communities and government agencies at local levels and turning the speculative concept of open data-driven governance into an efficient e-participation platform.

8 Discussion: Challenges of the Open Data-Driven Movement and Recommendations

8.1 The Economic Interests of Public Stakeholders and the Hidden Cost of Open Data

One of the most anticipated and often cited potential institutional challenges that the open data movement could face and somehow slow down its diffusion in Finland is that some public agencies could be reluctant to open their datasets since they might lose their revenues if they are obliged to publish their data on a free-of-charge basis (Lehtonen, 2011), which raises the concerns about the economic sustainability of a number of government services and bodies that have demonstrated impressive results during last decades in gathering information and building datasets on various sectors of public life and economy such as meteorological information, topographic and vector maps and other information. Some of the government information and datasets have been usually provided on a fee basis, contributing to the overall capability of public agencies to be at least partly self-reliant and less dependent on budget funding and support, which, as it turns out, now should be increased at the expense of taxpayers to negate the economic effects of the hidden cost in the free distribution of open data and ensure the continuous gathering, aggregation, preprocessing, sorting and, eventually, publication of such datasets without losing significantly in terms of quality and volume of generated public datasets.

Policy Recommendation: The Promotion of Fee-Based Distribution of Certain Government Datasets for Corporate Actors
One of possible solutions that could help to address the challenge associated with the financial unsustainability of current economic models, which often implies a free publication of open data by all public and semi-public agencies, is the provision of some government datasets exclusively on a commercial basis. This measure could potentially be applied to datasets, especially statistical information and meteorological information, that are currently provided in a systematic and, more importantly, automatic manner to various private organizations and transnational corporations. Another source of income could come from the delivery of large volumes of datasets and hard copies of government information in paper formats by ordinary post, which could hypothetically help to generate some revenues from postages, putting advertisements into the datasets and so on. For example, small- and mid-size businesses could receive a limited number of datasets free of charge. Larger business entities and transnational corporations that usually demand high volumes of data for their multiple analytical departments, market research surveys and an extended range of commercial products should pay a small fee to access such information without any limitations. Annual subscription could be the most feasible way to provide a sustainable and systematic flow of additional financial assistance to such public or semi-public open data depositaries. In this respect, tax deductions and charity assistance could also be regarded as a potentially cost-effective way to motivate local businesses and young technological firms to experiment with various options in the area, too. Naturally, any promising projects of political and socioeconomic importance for local communities, which are usually supported by non-governmental organizations, academic institutions and independent developers, should receive significant discounts or excluded from any payments whatsoever.

8.2 The Reuse of Open Data Requires Technological Skills

The public value of open data is that independent developers could reuse it without any copyright restrictions. The reuse, that is, processing of open data for developing third-party applications and projects, requires skilled enthusiasts and technologically savvy citizens who should be ready to contribute their time, knowledge and expertise (Akman et al., 2021; Hjalmarsson et al., 2015; Kassen, 2016; Sobaci & Hatipoğlu, 2017) to the creation or co-creation of open data-driven products often on a voluntary and free-of-charge basis. The key question here is what could motivate them to join the movement. Government agencies and non-governmental players not always could provide compensation for the efforts due to shortages of financial resources. In addition, the widely used practices to boost public sector innovations and activity in the area such as various contests, hackathons, competitions and so on, which usually promise prizes and other financial and non-financial motivations for potential developers to join the movement, and in particular participate and create open data-driven applications are only good in locating the best solutions and providing initial support for promising startups but should not be regarded as a sustainable way of the concept promotion on a more systematic and long-term basis. New implementation models of the concept promotion in private and non-governmental sectors should be proposed to mitigate the overall cost of the research and development in the area.

Policy Recommendation: Developing Open Services and Proposing Compensatory Funding Mechanisms for Independent Open Data Developers

Competition and commercialization should be one of the drivers to advance open data as a sustainable model of creativity in the area. One of possible solutions could be found in the development of business models that would compensate professional software developers for their efforts through such mechanisms as grants and public patronage, similar to a quite successful practice that is used today in scientific communities and academia to boost public sector innovations and new knowledge when researchers have to compete with each other for limited resources, which are usually provided by national or local governments and less often by private corporations, non-governmental organizations, development funds and charities. Another solution could be found in the development of universal platforms or standardized semi-finished products by professional developers on a fee basis, which could be then purchased by public or private stakeholders in order to ensure the economic sustainability of the open data-driven model through a comprehensive instrument of compensation of middlemen and make it freely accessible by everyone ***without any copyright restrictions*** as a new type of information, which could be tentatively called as ***open services***. Like any ordinary open data it could be reused or even modified by everyone without any physical or legal

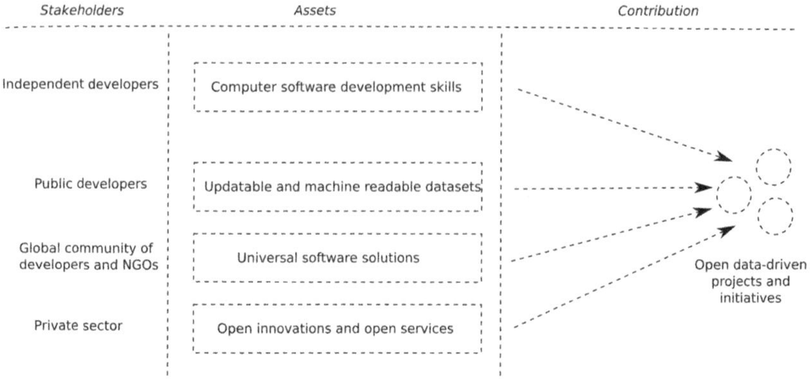

Fig. 4.5 The open data movement as a collaborative phenomenon. *Source: own illustration*

restrictions, presumably creating a self-sustainable mechanism of the open data movement. Global cooperation and networking should be advanced by non-government players in the area. There are some projects that prove to be effective through such standardization and openness in the distribution of universal software platforms or separate software codes and packages such as the CKAN, an open-source data management information system, which is being developed and supported actively by various public, private and non-governmental actors from different parts of the world, for example, from such countries as the United States, Canada, Australia, Mexico, Switzerland, Germany and United Kingdom (CKAN project, 2020). All these approaches and solutions could help to promote the open data movement as a truly collaborative phenomenon among its key stakeholders (see Fig. 4.5).

9 Summary: Key Findings

This chapter of the book claims to be an agenda-setting research, which was, first of all, aimed to explore a truly multidimensional nature of open data governance from the perspective of such a technologically developed nation as Finland. One of key findings of the research is that the open data concept could provide a promising collaborative platform to network various stakeholders in the area and accelerate public sector reforms. In this regard, it is difficult to underestimate the important role of public-private partnerships between key players of the open data movement, namely government agents, citizens, developers, non-governmental organizations, mass media and businesses. As the Finnish case demonstrates the successful realization of the idea demands a fairly equal contribution from representatives of both public and private sectors, whether it is in opening official government datasets, diversifying funding options, outreaching the related open data-driven activities in civil society, sharing knowledge and technologies or networking professionally with other peers so that to launch more or less sustainable open data-driven projects and startups in the area. Moreover, the active propagation of political and socioeconomic benefits of the concept, especially of those that are related to the processing and further reuse of open data in various third-party projects and collaborative platforms, was one of the most important aspects of the whole process. Another finding of the research was that the active promotion of the concept among developers leads to the creation of the related market niches where a new phenomenon of the open data-driven entrepreneurial activity of developers and business intermediaries emerges, which could have a profound multiplicative effect on the overall development of the idea. Finally, the case of Finland has also clearly pointed to the importance of cooperation with the local non-governmental sector and data journalists, whose active participation is crucial for the overall progress of the open data movement.

References

Ackerman, J. M., & Sandoval-Ballesteros, I. E. (2006). The global explosion of freedom of information laws. *Admin. L. Rev., 58*, 85.

Agience. (2020). http://agience.fi/

Akman, Ç., Akman, E., & Okcu, M. (2021). Turkey's open data policy. In *Handbook of research on global challenges for improving public services and government operations* (pp. 13–35). IGI Global.

Apps4Finland. (2016). http://www.apps4finland.fi/

Avoin Ministerio. (2020). www.avoinministerio.fi

Avoindata. (2021). The national open data project of Finland. https://www.avoindata.fi/en

Bengtsson, Å., & Mattila, M. (2009). Direct democracy and its critics: Support for direct democracy and 'stealth' democracy in Finland. *West European Politics, 32*(5), 1031–1048.

Berliner, D. (2010, August). Veto Players, Democracy, and the Diffusion of Freedom of Information Laws. In APSA 2010 Annual Meeting Paper.

CKAN project. (2020). The universal collaborative open source solution for data management. https://ckan.org/

De Smaele, H. (2007). Who's in control? Access to information as a crucial element in the balance of power between media and politics. https://lirias.kuleuven.be/handle/123456789/406063

Datademo. (2020). The participatory budgeting platform. http://datademo.fi/english/

FDPDS. (2020). The Finnish Digital and Population Data Services Agency. https://www.avoindata.fi/en

FMI. (2020). The Finnish Meteorological Institute. http://en.ilmatieteenlaitos.fi

Forum Virium. (2015). The Finnish champion of open data reveals the EU's biggest polluters. http://www.forumvirium.fi/en/press-release/the-finnish-champion-of-open-data-reveals-the-eu%E2%80%99s-biggest-polluters

Forum Virium. (2020). The Finnish champion of open data reveals the EU's biggest polluters. http://www.forumvirium.fi/en

Global Open Data Index. (2019). http://index.okfn.org/

Hielkema, H., & Hongisto, P. (2013). Developing the Helsinki smart city: The role of competitions for open data applications. *Journal of the Knowledge Economy, 4*(2), 190–204.

Hjalmarsson, A., Johansson, N., & Rudmark, D. (2015, January). Mind the gap: exploring stakeholders' value with open data assessment. In 2015 48th Hawaii international conference on system sciences (pp. 1314–1323). IEEE.

HLD. (2020). The Helsinki Loves Developers Project. http://dev.hel.fi/

HRI. (2020). The Helsinki Region Infoshare. http://www.hri.fi/en/

Hyvönen, E. (2020). Linked Open Data Infrastructure for Digital Humanities in Finland. In Proceedings of Digital Humanities in Nordic Countries (DHN) (pp. 254–259).

Jaakkola, H., Mäkinen, T., & Eteläaho, A. (2014, June). Open data: opportunities and challenges. In Proceedings of the 15th International Conference on Computer Systems and Technologies (pp. 25–39).

Jyväskylä Avoin Data. (2020). http://data.jyvaskyla.fi/

Kassen, M. (2015). E-government in Kazakhstan: The unitary model of implementation. In *Understanding systems of e-government* (pp. 74–131). New York, NY.

Kassen, M. (2016). Open government in Kazakhstan: A tentative touch of participative democracy? In *E-government in Kazakhstan* (pp. 66–83). Routledge.

Kassen, M. (2018a). Open data and its intermediaries: A cross-country perspective on participatory movement among independent developers. *Knowledge Management Research & Practice, 16*(3), 327–342.

Kassen, M. (2018b). Adopting and managing open data: Stakeholder perspectives, challenges and policy recommendations. *Aslib Journal of Information Management, 70*(5), 518–537.

Kassen, M. (2019). *Open data politics: A case study on Estonia and Kazakhstan*. Springer.

Kettunen, P. (2001). The Nordic welfare state in Finland. *Scandinavian Journal of History, 26*(3), 225–247.

Kuutti, H., Lauk, E., Nevalainen, P., & Sokka, R. (2012). Finnish media policy: Less restrictive, more directive. In *Understanding media policies* (pp. 100–115). Palgrave Macmillan.

Lehtomäki, J., Tuominen, S., Toivonen, T., & Leinonen, A. (2015). What data to use for forest conservation planning? A comparison of coarse open and detailed proprietary forest inventory data in Finland. *PLoS One, 10*(8), e0135926.

Lehtonen, P. (2011). Open data in Finland-Public sector perspectives on open data. Next Media, a Tivit programme.

MFF. (2020). The Ministry of Finance of Finland. http://vm.fi/en/frontpage

Mindtrek. (2020). http://www.mindtrek.org/

MJF. (2020). The Ministry of Justice of Finland. http://oikeusministerio.fi/en/

MTC. (2020). The Ministry of Transport and Communications of Finland. http://www.lvm.fi/home

NLOF. (2020). The National Land Office of Finland. http://www.maanmittauslaitos.fi/en

ODP. (2020). Open Data Program of Finland. The Finnish Ministry of Finances. http://vm.fi/documents/10623/364262/Open+data+programme_broschure/e65d416e-7d06-449a-9560-745560df048b

OECD. (2019). Government at a Glance 2019: Finland. http://www.oecd.org/finland/govataglance.htm

OFC. (2020). The Open Finland Challenge. http://openfinlandchallenge.fi/en/

OGP. (2013). The Open Government Partnership. http://www.opengovpartnership.org/country/finland

OKF. (2020). The Open Knowledge Finland. http://fi.okfn.org/about/

Ominvoimin. (2015). The peer-to-peer innovations generating community platform. http://ominvoimin.com

Oulu Avoin Data. (2020). http://www.ouka.fi/avoindata

Päätökset. (2020). The government decision-making tracking project. http://dev.hel.fi/paatokset/

Perkmann, M., & Schildt, H. (2015). Open data partnerships between firms and universities: The role of boundary organizations. *Research Policy, 44*(5), 1133–1143.

Protomo. (2018). Business innovations networking platform. www.protomo.fi

Pynnönen, S., Haltia, E., & Hujala, T. (2021). Digital forest information platform as service innovation: Finnish metsaan.Fi service use, users and utilisation. *Forest Policy and Economics, 125*, 102404.

Saario, V., & Ginsburg, R. B. (1965). International co-operation in litigation: Finland. In *International co-operation in litigation: Europe* (pp. 105–118). Springer.

Shiramatsu, S., Tossavainen, T., Ozono, T., & Shintani, T. (2015). Towards continuous collaboration on civic tech projects: Use cases of a goal sharing system based on linked open data. In *Electronic participation* (pp. 81–92). Springer International Publishing.

Sihvonen, M., & Sipilä, M. (2014, May). User-Based Open Knowledge Increasing E-Democracy in Finland. In Conference for E-Democracy and Open Government (p. 509).

Sitra. (2020). The Finnish Innovation Fund. http://www.sitra.fi/en/about-sitra

Sobaci, M. Z., & Hatipoğlu, İ. (2017, May). Measuring the engagement level of political parties with public on Facebook: The case of Turkey. In 2017 Conference for E-Democracy and Open Government (CeDEM) (pp. 209–216). IEEE.

Statistics Finland. (2020). http://www.stat.fi/index_en.html

Still, K., Huhtamäki, J., Russell, M. G., Basole, R. C., Salonen, J., & Rubens, N. (2013, January). Networks of innovation relationships: Multiscopic views on Finland. In ISPIM Conference Proceedings (p. 1). The International Society for Professional Innovation Management (ISPIM).

Tampere Avoin Data. (2020). The open data project of the city of Tampere. http://www.tampere.fi/tampereen-kaupunki/tietoa-tampereesta/avoin-data.html

The 6Aika project. (2020). http://6aika.fi/in-english/

The Cloud'N'Sci.fi project. (2020). https://cloudnsci.fi/

The Parliamentary Explorer project. (2020). https://www.microsoft.com/en-us/store/apps/eduskunta-explorer/9nblggh085ww

The puoluekentta project. (2020). http://puoluekentta.tstm.info/

The rOpenGov initiative. (2020). The data-driven collaborating and networking platforms for software developers. http://ropengov.org/

Timonen, V. (2003). *Restructuring the welfare state: Globalization and social policy reform in Finland and Sweden*. Edward Elgar Publishing.

Tredea. (2020). Tampere Region Economic Development Agency. http://www.tredea.fi/en/tredea

Ultrahack. (2020). https://ultrahack.org/

UN. (2020). United Nations E-Government Survey 2020. https://www.un.org/development/desa/publications/publication/2020-united-nations-e-government-survey

Valtori. (2020). The Valtori Government ICT Center. http://www.valtori.fi/en-US

WJP OGI. (2020). World Justice Project – Open Government Index. https://worldjusticeproject.org/rule-of-law-index/country/2020

Ylipulli, J., & Luusua, A. (2020). Smart cities with a Nordic twist? Public sector digitalization in Finnish data-rich cities. *Telematics and Informatics, 55*, 101457.

CHAPTER 5

Open Data Governance in Sweden: Government Data Transparency in the Context of Social Democracy

Abstract In 2021 Sweden celebrated the 255th year of the first freedom of information law in the history of humankind, that is, the first attempt of citizens to promote the transparency of government institutions through the release of public information. In this connection, this research presents a case study that investigates open data governance in the context of a country that is universally well-known today not only as a historical homeland of such legislation but also as a global pioneer of the open knowledge movement that explicitly propagates the public value of free access to public information. The development of this extremely liberal information policy has been mirrored recently in the widespread diffusion of other related promising concepts such as open data, open economy, open science, open state, open source software, net neutrality and even pirate politics. In this regard, by adopting a wide range of open data-driven projects and startups, Sweden not only has followed the global open data movement but, more importantly, managed to promote the concept further, offering a new theoretical foundation to reincarnate the good old idea of e-government from a somewhat notorious cliché of top-down and purely bureaucratic tool of public sector reforms into a promising ideological platform to advance civic engagement, e-participation, freedom of press and even social democracy in a completely new innovative manner.

M. Kassen, *Open Data Governance and Its Actors*, Studies in National Governance and Emerging Technologies,
https://doi.org/10.1007/978-3-030-92065-4_5

Keywords Open data • Sweden • Social democracy • E-participation • Open knowledge

1 Setting the Research Agenda

1.1 The Purpose of the Research

The open data movement is beginning to play an ever-increasing role in public sector reforms all over the world. Likewise, the research on open data as a theoretical concept is on the rise in many fields. However, the related literature rarely focuses on the analysis of empirical evidence from real practices and even less from insider multiple stakeholder perspectives in countries that are leading the trend in a global sense. In this regard, the main purpose of the research in this chapter is to study a multidimensional nature of open data in an unusual context of the country that has been historically famous as a homeland of the oldest freedom of information legislation, which is also universally well-known today as a global ideological leader of such promising concepts, related to the philosophy of open data, as open economy, open science, open state, open source software, net neutrality and even pirate politics in an attempt to understand the influence of socioeconomic and, more importantly, political backgrounds on the development of open data phenomena. This research presents a case study that investigates open data from multidimensional contexts of the country background in political, economic and social areas that could help to better understand the promising potential of the open data concept to enhance civic engagement and digital democracy and analyze more in detail the fundamental ideological factors of the open data movement as a collaborative political phenomenon.

1.2 Choosing a Case for the Research

Taking into account an intrinsically collaborative nature of its implementation, which as a rule requires transparency of public sector, strong political will and, more importantly, readiness of local population to resort to the promise of e-participation and civic engagement, open data as a concept highly depends on country contexts in various areas. In this regard, the advent of open data governance as a political phenomenon shortly after the adoption of the famous open government directive by the Obama administration in early 2009, which unambiguously declared the ideals of transparency, participation and collaboration as hallmarks of open society, found a fertile field in Sweden. Being globally well-known as the first nation that as early as 1766 introduced the freedom of information law in the history of humankind (Nordenstreng, 2017; Nordin, 2016) and universally famous for its long-standing adherence to traditions of liberalism, social justice and direct democracy in many areas of public policy, this Scandinavian nation could be regarded as one of the most ideal places to promote the concept of open data as an important part of this new political philosophy of governance. In actively propagating the ideals of secularism, pluralism and voluntarism in political and social arenas, defending the public value of social democracy and open competition in the economic area, advancing the concept of multiculturalism, open society and free education in the social area or embracing a truly altruistic philosophy of open source software and open technological platforms in the telecommunication field, Sweden provides favorable political and socioeconomic contexts to promote open data as a multidimensional socioeconomic phenomenon.

Moreover, the existence of technologically developed and industrially diverse local economy and social welfare state as well as the presence of a very active civil society and non-governmental sector that values an innovative approach and proactive stance in furthering digital progress and technological knowledge (Larsson & Malmberg, 1999; Wallmark & McQueen, 1991) has created a really favorable environment to aggregate an extremely wide range of datasets and statistics that have previously been compiled for decades in various formats at many institutional levels. These stockpiles of socio econometric data present almost an unlimited source of government information that could be untapped now and used by members of local communities in creating new third-party open data-driven products and even peer-to-peer services. In this regard, the emergence of a plethora of open data-driven projects and platforms in various areas of the economy at both national and local levels of governance (Grundstrom & Lövnord, 2014) helps this Nordic nation to boost open innovations, especially by making its public sector reforms presumably more cost-effective and result-oriented. In this regard, it seems especially interesting to investigate the development of such a new concept as open data governance, which realization, according to former US President Barack Obama (2009), should ideally be based on three classic principles of public openness, namely transparency, collaboration and participation, in the context of the country that has been universally well-known for its centuries-long traditions of open data philosophy. This philosophy is reflected today in the active promotion of such related concepts as freedom of information, open state, open economy, open society, open software and open science. It is this political reincarnation of open e-government policies reflected in the recent emergence of the open data concept and traditions of freedom and openness, which is deeply embedded in the Swedish society, that create an ideal venue to promote civic engagement as a modern tool of public sector reforms and advance the open data movement as a political philosophy (see Fig. 5.1).

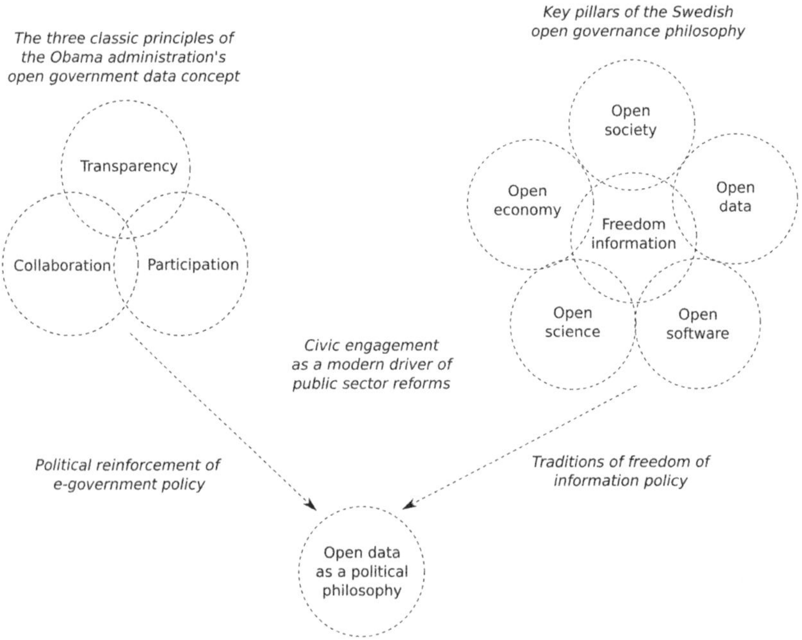

Fig. 5.1 The development of the open data movement as part of a larger political ideology of freedom of information philosophy in Sweden. (Source: Own illustration)

1.3 Understanding the Academic Background

There are many case studies in academic literature that research the open data governance in various countries, including in Sweden (Axelsson & Schroeder, 2009; Elder, 2013; Janssen, 2012; Lassinantti, 2014; Lassinantti et al., 2014). Analyzing the phenomena, researchers usually focus on studying the innovative aspects and economic implications of the concept development, such as the increased level of communication among its main stakeholders (Lassinantti et al., 2014), institutional dimensions of how the phenomena are promoted in governance (Safarov, 2019), especially at local levels (Susha et al., 2015), public sector innovations (Rudmark & Hjalmarsson-Jordanius, 2019), e-climate (Klein et al., 2012), e-health platforms (Cool, 2015; Timpka et al., 2014), cost-effective business and technological models in data management (Stråle & Lindén, 2014) or investigating the public value of open data journalism (Appelgren & Nygren, 2014; Widholm & Appelgren, 2020) and so on. As it is demonstrated in the topic of previous research in the area, open data is truly a multidimensional phenomenon that helps to promote a wide variety of innovations in various sectors of economy and in a range of public sector reforms. In this regard, it is interesting to focus on the analysis of new collaborative and networking processes that are beginning to be promoted between key stakeholders of the open data movement in various political and socioeconomic dimensions of these reforms. As such, this research presents a case study that investigates the promising potential of the open data concept to enhance public altruism, philanthropic activity and freedom of information in a more holistic manner. This research also aims to understand better the fundamental factors of the open data movement as a collaborative multidimensional political phenomenon by resorting to an interesting case of a country that is universally regarded today as one of the global leaders in open data technologies, according to various international rankings (Global Open Data Index, 2019; OECD, 2019; UN, 2020). For example, in 2020, Sweden was ranked first among all nations in a prestigious global open data index (WJP OGI, 2020) (see Table 5.1).

2 The Key Tools of the Analysis

The truly rapid and widespread diffusion of open data phenomena all over the world and, more importantly, an interesting trend in recent emergence of new promising open data-driven ideological platforms to advance

Table 5.1 The Global Open Government Index in 2020

Country	*Ranking*
Norway	1
Denmark	2
Sweden	3
Finland	4
Netherlands	5
New Zealand	6
Australia	7

Source: Sample data from the Global Open Government Index (WJP OGI, 2020)

political communication among various stakeholders, both public and private, around the open data movement in Sweden is ripe for close investigation and policy analysis through the prism of actual cases in a real-life context. In this regard, this phenomenon will be explored through the prism of its key actors and stakeholders, identifying the main driving forces and challenges of the open data movement with the primary goal of the whole case study is to understand how this promising idea of open governance is developing in an unusual context of social democracy in a country that is also globally well-known for its excellent sophisticated solutions in the area of digital governance, especially demonstrating really impressive results in promoting open data-driven platforms. The context analysis of the country background will be helpful in studying the political and socio-economic environments that are hypothetically conducive to the development of the concept and assessing the readiness of key stakeholders to adopt and lead the open data movement. The next research method, a policy analysis, is based on the close investigation of realization and regulation strategies adopted by the Swedish government mostly at the central level as well as the observation of interesting patterns reflected in the emergence of new trends in political communication processes among key policymakers of the open data movement, which could help to shed light on the role of private and non-governmental sectors in the diffusion of the concept in this Scandinavian country and reveal related networking activities among various stakeholders, especially developers, journalists, non-governmental organizations, academia, businesses and technically savvy citizens. Finally, the content analysis of open data-driven projects launched independently by the members of civil society could help to understand the civic component of the open data paradigm. In general, the combination of these three methods will be useful in understanding better political, social and economic implications of various country contexts on the development of the open data movement, locating the most important institutional drivers and challenges that the policymakers face today in the realization of related digital collaborative and participatory platforms and offer a number of practical recommendations on how to overcome them.

3 The Country Context Analysis: Understanding Political and Socioeconomic Backgrounds of the Open Data Movement in Sweden

3.1 The Political Contexts: Individual Freedoms and Transparency of Government as Absolute Public Values

Sweden is universally well-known as one of the most transparent and democratic societies in the world with really long-standing centuries-old traditions of civic engagement and direct participation of citizens in political decision-making at both national and local levels of government, which resulted in the creation of a truly collaborative, egalitarian and socially responsive public sector. Public trust in government and capacity of government institutions to advance the nation is a distinctive feature of Swedish society. Individual freedoms of commons have always been a remarkable feature of the Swedish society throughout its national history. Due to the cold climate conditions, historically, *manorialism* and *feudalism* as socioeconomic phenomena have never played an important role in the development of this Nordic country during the medieval times as it had in other parts of the European continent (Heckscher & Heckscher, 1954; Trägårdh, 2007). The relationship between seniority and commons has never been antagonized as a conflict between social strata. These traditions of individual freedom have eventually cemented the unique Scandinavian public mindset of an egalitarian society. The universal respect of such key public values as equality and social justice has helped to build one of the most open societies in the world, with a really strong representation of civic communities in local governance. All these prominent features of political and historical backgrounds have helped this nation to progress economically and socially alike and now provide a favorable environment to promote the open data concept as a modern ideological platform to advance democracy and civil society in a new digital context by embracing and furthering the main public values of the concept, that is, transparency, participation and collaboration.

Political transparency and traditions of civic participation in governance allow various political groups in Sweden to create highly liberal political movements. One of the unique trends is a promotion of such political ideology as pirate politics (Burkart, 2014; Dvorak, 2006; Erlingsson & Persson, 2011; Jääsaari & Hildén, 2015; Li, 2009; Miegel & Olsson, 2008), which propagates such key values of political and socioeconomic digitalization as liberalization of copyright laws, open source society, e-participation, e-democracy, open knowledge, open science, open data, freedom of information and net neutrality. This trend has been recently reflected in the promotion of the Pirate Party in Sweden. In this regard, the notorious scandals with WikiLeaks have partly popularized the pirate politics ideologically in mass media and social networks (Lindgren & Lundström, 2011; Steinmetz, 2012). Although the party has never won any seats in the national parliament, its emergence nevertheless as a political organization with a concrete ideological agenda has activated political debates on these topics of freedom of information and access to government files and records at national and international levels, creating a very favorable political and media entourage around the movement that eventually helped to advance the open data movement, which basic pillars really closely correlate with the main principles of this unorthodox ideological platform.

3.2 The Economic Context: Open Data-Driven Economy as a Platform to Boost Public Sector Innovations

Sweden has a technologically advanced economy with strong positions in ICTs. The deregulation in the telecommunication area has always been strongly supported by the national government and local companies in the area (Berggren & Laestadius, 2003; Eriksson et al., 2019; Henrekson & Jakobsson, 2005; Hultkrantz, 2002). This policy allowed to create a liberal legal and socioeconomic environment to advance open innovations indirectly in various sectors of economy. In this regard, such internationally recognized and successful Swedish hi-tech and industrial companies as, for example, Ericsson, Volvo, TeliaSonera, Klarna, Electrolux, Sandvik, Nordea, Symsoft, Dorotelecoms, Tele2 AB, Scania, Handheld Group, Securitas, SAS Group and many other actors have contributed greatly to the creation of an extremely competitive market that relies heavily on research and development as a strategy of sustainable economic growth. The telecommunication market that values innovations and cost-effectiveness in many areas of economic activity (Andersson et al., 2005; Blomstrom & Kokko, 2003; Khayyat, 2017) creates a favorable atmosphere to promote various technological projects whether it is in the production of modern telecommunication equipment, advancement of e-commerce and e-trade platforms or development of business solutions based on the use of open source software. Moreover, the spirit of entrepreneurship and economic freedom that coexists in harmony with a really popular sense of altruism and readiness to help local communities to prosper and thrive (Lundström, 1996; Vamstad & von Essen, 2012) that is deeply shared and cultivated by local business communities and corporations (Lundqvist & Williams Middleton, 2010; Mecherkany, 2010) is really conducive for the emergence of various collaborative public-private partnerships that support an extremely wide range of social projects and startups aimed at developing local neighborhoods, including in the open data area (see Fig. 5.2).

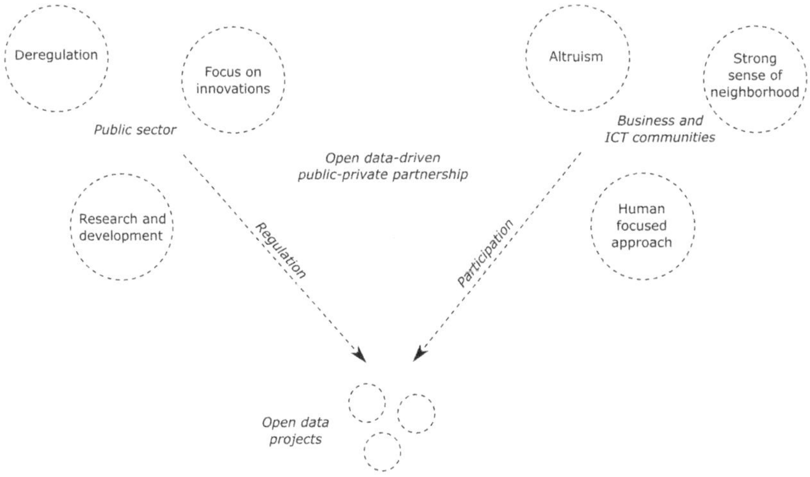

Fig. 5.2 The fundamentals of the open data-driven public-private partnership in Sweden. (Source: Own illustration)

3.3 *The Social Context: Open Society as a Fundamental Factor of Civic Engagement and Participation*

Open data is a context-dependent phenomenon. In this regard, the adherence to the Scandinavian model of social democracy has resulted in the creation of favorable conditions for the advancement and propagation of open data among various civic stakeholders (Joshi & Navlakha, 2010; Juell-Skielse et al., 2014). As one of the nations that support various programs of social welfare for all residents whether it is in providing free education, universal healthcare or extensive social security programs, Sweden has managed to create one of the most egalitarian societies in the world where a sense of social equality and philanthropy are universally regarded as key pillars of public mindset. In this regard, this Scandinavian nation could be proud to be one of the most educated and liberal countries in the world with a really widespread popularity of such liberal values and ideas in the economic, academic and ICT areas as, for example, open politics, open innovations, open economy, open governance, open source software, open science, open journalism, open knowledge, open market and so on. The propagation of all these ideas creates a truly friendly political and socioeconomic atmosphere to promote the open data concept, too. For example, a very strong and influential community of local developers and technology enthusiasts truly embraces and praises the philosophy of open source solutions in the sphere, providing an additional stimulus for the advancement of open data in various third-party applications. One of the main ingredients in the successful promotion of the concept presumably lies in the ability of local developers to network and support independent and, more importantly, non-profit open data-driven projects. Historically, the strong traditions of self-governance, individual autonomy and voluntarism among citizenry have provided a really fruitful soil to activate civic engagement and e-participation in the area, helping the society advance further in building digital democracy.

3.4 *The International Contexts: Being Globally at the Forefront of the Open Data Movement*

Sweden was among the first nations in the world that joined the Open Government Partnership shortly after its creation in 2011 and has been committing ever since to promote its key overarching components (OGP, 2015). The promotion of various open data platforms is one of the most promising trends in the area since both public and non-governmental stakeholders are engaged in the area. For example, Sweden is one of the globally well-known proponents of the humanitarian open aid movement. Being as one of the most active promoters of peace-keeping operations and humanitarian foreign assistance programs around the world, this country spends a significant amount of its gross domestic product (GDP) for international projects in order to help numerous countries in the developing world, for example, in fighting poverty, diversifying industries, building new roads, schools and providing shelters for affected populations and feeding refugees in various humanitarian camps across the world. In this regard, the decision of the national government to open all data that is related to the topic from all public resources, which are engaged in humanitarian operations, in the project named Open Aid helps to promote efficiency in the area. As the recent studies shows, these transparency measures could potentially help to accelerate public support and ensure better public trust for the continuation and enhancement of related relief projects and programs (Linders, 2013; Olafsson, 2012; Weaver et al., 2014), increasing public knowledge among both domestic and international audiences, especially non-governmental organizations and charity foundations about a really diverse range of humanitarian operations provided every year by this nation and, more importantly, helping to better understand a positive socioeconomic impact that such aid may have in different parts of the world in easing the plight of affected population.

4 Open Data Regulation: Being a Long-Standing Promoter of the Freedom of Information Philosophy

The respect for the principles of open information requests (*Offentlighetsprincipsbegäran*) is a key pillar of open data governance in Sweden. For example, such an open data-driven initiative as a Handlingar project provides an opportunity for civic actors to make freedom of information requests or browse requests to government by other citizens (Handlingar project, 2020). Being universally well-known as the first nation that introduced the freedom of information law in the history of humankind as early as 1766 (Ackerman & Sandoval-Ballesteros, 2006; Fink, 2018; Mustonen, 2006; Rydholm, 2013), Sweden could be regarded as one of the global pioneers in the promotion of the concept as a legal and, more importantly, political platform that guarantees the citizen's right to monitor the activity of the public sector in various spheres of national economy and advancing the development of civil society institutions. The Swedish Age of Liberty or *Frihetstiden* (1719–1772), which was reflected in the adoption of a number of liberal acts and proclamations by the Swedish Parliament, helped to establish such basic principles of open society as transparency of public records and free access to government information. This eventually helped to build quite favorable historical political and socioeconomic conditions to actively promote today a wide range of e-participation and civic engagement projects aimed at furthering the openness of government but now in a digital context, too. In this regard, the celebration of the 250-year anniversary of the Swedish FOIA in 2016 and related events organized by the UNESCO provides an additional media and political incentive to promote open data and develop further related legislation as a platform to advance the concept not only as a reactive tool in disclosing the government information and datasets reflected in the traditional FOIA laws, such as The Swedish Personal Information Act (1973), but also as an ideological instrument to further a new mechanism of open data governance that helps to publish *proactively* all information and datasets in special depositories for further reuse by all members of civil society (see Fig. 5.3).

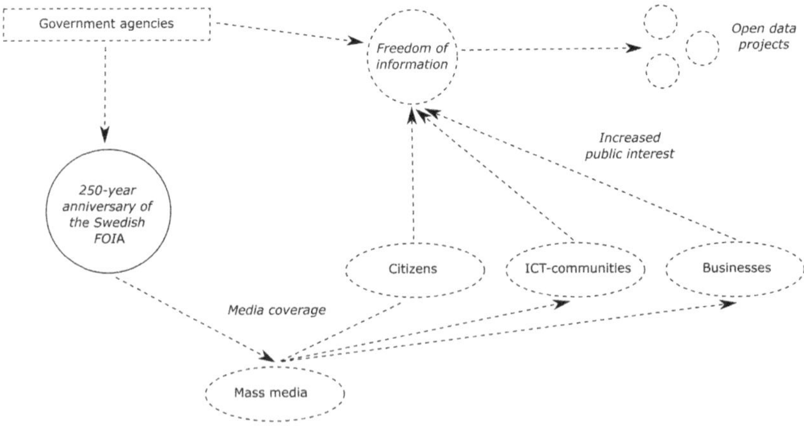

Fig. 5.3 The 250-year anniversary of the Swedish FOIA provided an additional political incentive to promote the open data-driven philosophy in governance in 2016. (Source: Own illustration)

5 The Stakeholder and Policy Analysis of the Open Data Movement: Stakeholders and Realization

5.1 *Government Agencies: Propagating the Public Value of Open Data Philosophy*

The role of the public sector in the advancement of the open data concept in Sweden is truly crucial. For example, in order to advance innovations in government, including through open data, a special government body was created, *the Agency for Digital Government* (Myndigheten för digital förvaltning). This and other government organizations are key ideological promoters of the open data governance concept at both national and local levels of public administration even despite the fact that it is a unitary state where many public projects in the area are initiated, funded and realized, first of all, by central authorities, which, in combination with existing single legislation, single e-government networks and single databases, provides a really effective instrument to implement various national digital platforms in the e-government area. In this regard, such public agencies as, for example, the Ministry of Enterprise and Innovations, the Government Agency for Innovation systems (VINNOVA), the Swedish Statistics Agency, the National Archives (*Riksarkivet*), the Swedish Agency for Growth Policy Analysis, the National Heritage Board, The Swedish Mapping, Cadastral and Land Registration Authority (*Lantmäteriet*), the Swedish International Development Agency (SIDA), the Ministry of Health and Social Affairs, Swedish Companies Registration Office (*Bolagsverket*) and many other government stakeholders play an important role in the promotion of the open data concept and regard it as, first of all, a cost-effective platform to further the public sector innovations and reforms. All these government agencies not only publish open data as a raw material designed to be used by third-party applications but also try to encourage local communities of developers to reuse the datasets by organizing various public relations, collaborative events and networking meetings such as hackathons, master classes and app contests. For example, the Hack for Sweden app contests are regularly organized to provide a new collaborative platform to further the idea of open data-driven communities, which use a combination of datasets from various public agencies and departments in an attempt to boost public sector innovations in such areas as advancing various solutions to improve working and living environment for local communities or businesses.

In this regard, founded in 2001, VINNOVA is a leading government agency that promotes research and development in various sectors of economy to boost innovations (Jacob, 2006) and supports various public projects (Chaminade & Edquist, 2006) in an extremely wide range of areas whether it is in building the national capacity to develop and adopt various open data technologies and platforms, creating an open business climate for foreign investors, promoting the international cooperation in the area or supporting various research projects in academic institutions throughout the country (Vinnova, 2020). The organization is also aimed to facilitate the mutual cooperation between government agencies, businesses and research institutions in finding cost-effective solutions for public and private sectors that could help the nation to further as a global pioneer of innovations and smart technologies in a number of industries. In this respect, one of the primary focuses of the agency is the development of conceptual frameworks to promote open data-driven innovations that could help advance the whole nation in this direction, first of all, by resorting to the untapped potential of the open data concept to further democratic processes in the public sector and create new market niches for developers and businesses, especially in e-commerce areas. The organization not only offers various funding options for contestants in the area at the national level but also helps talented innovators from Sweden to apply for the EU funding within the Horizon 2020 project in an attempt to promote collaboration and networking in research and development in a cross-border manner. Many technological parks and innovation hubs as well as universities and research institutes actively participate in these publicly supported projects and platforms.

5.2 *The Non-governmental Sector: Civil Society as a Main Ideological Fundament of the Open Data Movement*

The civic and non-governmental sectors play a crucial role in the diffusion of the open data movement in Sweden. The reuse of open datasets by civic actors helps to advance the concept as a collaborative and networking process where co-development and exchange of knowledge is a necessary prerequisite for the successful development and progress. For instance, the organization of Civic Tech Sweden workshops (CTSW, 2020) provides an interesting example of how such civic engagement could be accelerated in the area among technically savvy citizens. Such non-governmental organizations, independent networking platforms, research foundations, development groups and think tanks as, for example, The Engineers Without Borders, The Techno Creatives, Chalmers Ventures, Grant Thornton, Eudoxa, Timbro and many other stakeholders provide a wide range of services to civil society and public sector, offering expertise and consultations in building open economy and transparent government, including through propagating, supporting and advancing open data and related open innovations concepts and technological platforms in various sectors of economy. One of the interesting features in the development of the concept in Sweden is that not only non-governmental players and businesses participate actively in the promotion of open data but also cultural organizations. For example, various museums and exhibition centers are starting to publish scanned illustrations of famous paintings as open data. For example, the Swedish National Museum has already uploaded more than 6000 high resolution images of the most popular paintings as part of the Wikimedia Commons and through its open collection project (Nationalmuseet, 2021).

5.3 *Developers: Diffusing the Open Data-Driven Philosophy Among Innovators*

The ICT sphere could also be regarded as a certain driver of public sector innovations in Sweden in many senses whether it is in developing various e-democracy, e-participation, e-government, e-banking, e-commerce, e-trade or e-business solutions, which could be then commercialized, including in the global context. These innovations create a surplus value in producing technological products and services or realizing various digital projects that help to make the public sector more flexible, collaborative and responsive to the needs of local population in a context that has a multiplicative and longitude cost-effective effect on the development of the whole civil society. One of the most positive effects is reflected in the creation of a really sustainable local economy, higher level of quality of life and safer environment due to the automation of many administrative processes and transition to greener economy (Lfstedt, 2007; Ruuska & Teigland, 2009). In this regard, the existence of an army of professionals, representing both public and private stakeholders as well as independent professional networks of ICT specialists and even individual technically savvy citizens, creates a really favorable environment to promote the open data-driven philosophy since the successful development of such projects usually depends heavily on the participation from the part of local developers, designers and coders. Such collaborative and networking venues that are designed to boost co-development among local ICT experts with businesses, academia and with each other in a peer-to-peer manner as, for example, the Popdevelop project (PDP, 2017), the Startup Grind Stockholm (SGS, 2020), the Swedsoft Innovation Center (Swedsoft, 2020), TechMeetups Stockholm (TMS, 2020), the Open Data Meetups in Stockholm (ODMS, 2020), the Swedish Startup Space (SSS, 2017), the Swedish Software (SS, 2020), Dopter (Dopter, 2020), the Meta Solutions Project (MSP, 2017) and many other communication platforms help to promote the exchange of knowledge and expertise between all these stakeholders. These networking platforms also allow to increase dramatically the competitiveness of extensive ICT communities, indirectly indicating to the existence of a presumably closed cycle of open data-driven networking philosophy of open data governance where independent and professional developers from both private and corporate sectors of economy as well as technically savvy citizens contribute together to the advancement of data-driven innovations in various areas (see Fig. 5.4).

The social programs that provide free education for residents in local universities and the popularity of the ICT profession among youth offer great opportunities for the development of the open data concept in various economic niches. Furthermore, the business stakeholders who are interested in the advancement of related startups, especially in the relationship to the idea of big data as a trigger of open data-driven innovations in the commercial sector, pay special attention to the sphere of the national economy, funding or organizing numerous research projects and startups. For example, private venture funds, professional development networks, engineers groups and technological companies such as Sandvik, NetClean, Semcon, Stena, Sigma Technology, Volvo Group, Woman in Tech, Venture Cup and many other stakeholders help to advance innovations in the area, by supporting young startups or related networking and development projects in the area. The commercialization of open data phenomena into a profitable business model is one of the most popular ideas among local entrepreneurs and generally regarded by representatives of both government agencies and private corporations as an untapped source of innovations and economic growth. In this respect, many ICT companies in Sweden invest significantly in various innovative sectors, including in projects that are driven by big or open data. New business solutions based on

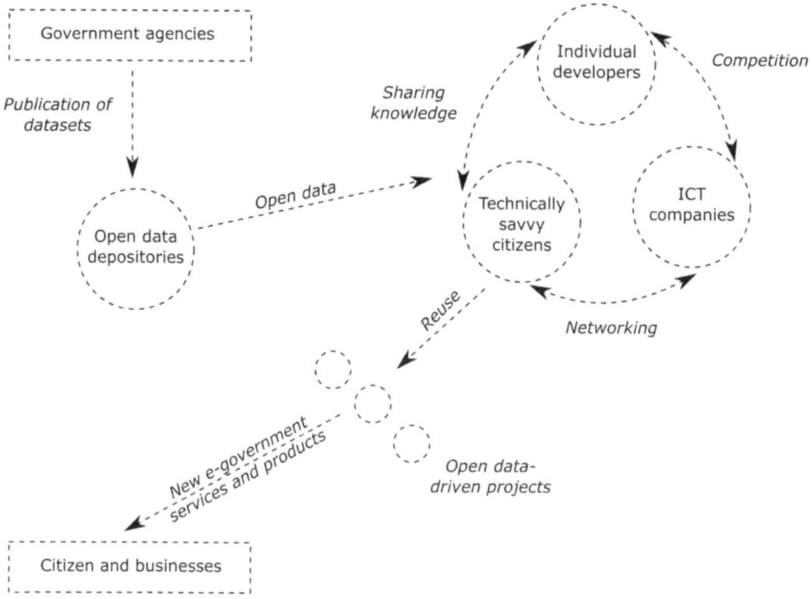

Fig. 5.4 The closed cycle of open data-driven networking philosophy. (Source: Own illustration)

further processing and reuse of various government datasets could offer sustainable and, more importantly, profitable economic models to innovations. By funding various research activities in the area in academic institutions and various think tanks in an attempt to boost technological and scientific expertise and raise public awareness, these private stakeholders support a huge army of developers, representing different professions in the area, starting from software engineers and designers and ending to data scientists and data journalists. Moreover, both public and private entities support the idea of app competitions as a practical way to promote the open data concept among independent developers by providing funding for the organization of various hackathons, app contests and professional awards in the area, for example, Hack the Crisis (2020), the Kid Hack Day (KHD, 2020), the Tele2 Hack Day (Tele2, 2017), the Health Hack Day (HHD, 2015), the Hackaway project (HAP, 2020), the Nordic Internet of Things Hackathon (NITH, 2020), the Hack for Sweden (HFS, 2020), the East Sweden Hack (ESH, 2020) and so on. These collaborative and networking events not only help to find the best open data-driven solutions for the needs of public sector but also encourage local computer experts and specialists to participate more actively in furthering the conceptual framework of open data-driven civic engagement and networking platforms in partnership with academic communities, taking into account the fact that there are a lot of potential undergraduate and postgraduate students among independent developers there.

5.4 *Academia: Developing the Conceptual Foundations of the Open Data-Driven Agenda*

Academic and research communities play an important role in advancing the open data movement in Sweden. In this regard, the research activity in such globally prominent venues of academic education and research as, for example, the Royal Institute of Technology, the Stockholm University, the Kista Science City (KSC, 2020), the Swedish Institute of Computer Science (SICS, 2020), the Lund University Internet Institute (LUII, 2020), the Research Institute Acreo (RIA, 2020), the Linköping University (LU, 2020) and many other educational and research institutes helps to boost innovations in the open data area (see Fig. 5.5). For example, the KSC is universally regarded as one of the global leaders in innovative research in telecommunications and computer technologies. Most of the research and development that happens in these areas is often done in close cooperation with both public and private stakeholders in an attempt to promote data-driven innovations in, for example, e-government or e-commerce. Another example of how public-private partnerships could be promoted in the area is an active cooperation of the many local education institutions on an important project called the Swedish National Data Service (2020) in sharing research data across universities, research institutes and laboratories in this country and beyond. Academic communities that participate in these projects directly benefit from networking and exchange of new data and research results that could be reused by peers across various professions and industries to boost open innovations and new data-driven research and educational products and standards, truly putting Sweden at the forefront of the global open science and open knowledge movements.

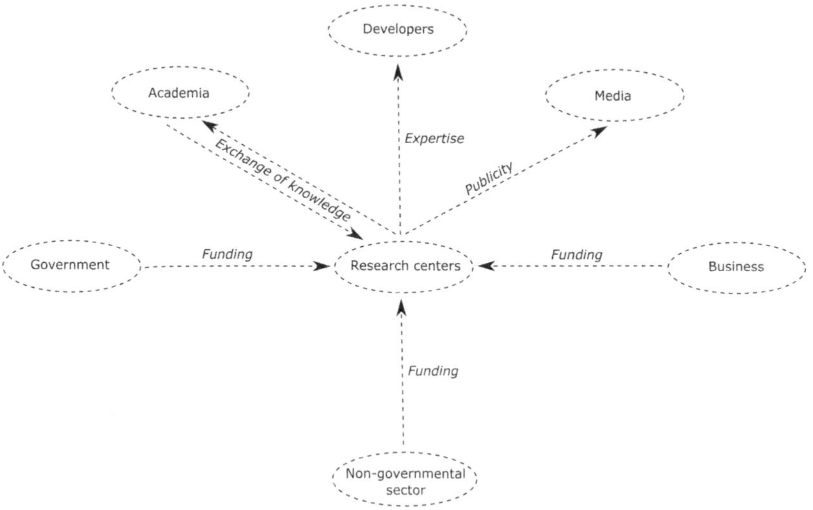

Fig. 5.5 The research centers as generators of knowledge and innovative ideas in the open data area. (Source: Own illustration)

5.5 *The Local Journalistic Communities: When Media Entourage and Publicity Really Matter*

The open data concept is actively covered and propagated in local media and social networks. Many media venues, both traditional and electronic, often publish analytical reports and articles on the topic, contributing greatly to the general increase of public knowledge about open data phenomena in this Scandinavian country. Furthermore, one of the growing networking communities among local media specialists is an active army of data journalists, who intrinsically rely on the analysis of various types of public and public datasets for their journalistic investigations. For example, open data-driven journalism, which has become popular in Sweden, was hallmarked recently in launching a special networking platform in the area named Journalism++ project (JPP, 2019). In this respect, open data journalism is widely understood among many media professionals as a new dimension of the profession that is intrinsically based on an active use of various public and semi-public datasets by journalistic communities in preparing their own media products and digital solutions. The open data phenomenon itself is regarded by local journalistic communities as a certain hallmark of new opportunities for a professional growth and development in the field, which could presumably boost greater responsibility, objectivity, democracy and greater transparency of many hidden political and socioeconomic processes, which could not be fully understood if journalists would only use traditional methods of investigation and analysis in the area. With the advent of data journalism, the media profession is enriched with more reliable and evidence-based ways to prepare journalistic materials.

6 The Overall Evaluation of Open Data Realization Policies in Sweden

The official open data project of Sweden (Oppnadata, 2020) is a classic example in the realization of the centralized digital government platforms that could be tentatively called e-centralism (Kassen, 2015), which as a rule hosts all datasets from various government agencies in one venue due to the unitary structure of the administrative and territorial division of the nation. However, even despite the fact that the national open data portal might be regarded as a number one project in the area, there are also several regional platforms that have been launched by local municipalities in such metropolitan areas as, for example, Stockholm (Open Stockholm, 2020), Lund (Open Lund, 2020), Malmö (Open Malmö, 2020), Umeå (Open Umeå, 2020), Uppsala (Open Uppsala, 2020) and many other cities and counties. Currently, the portal hosts more than 300 datasets in such categories as, for example, environment protection, science, politics, geography, transport, technology, law, agriculture and food and so on.

The World Justice Project puts Sweden third in its 2020 Global Open Government Index, which globally monitors the rule of law, including government transparency in accordance with the evaluation of national open data projects provided by e-government experts in the area, rating eight key pillars. The index ranks countries in accordance with the rating system ranging from 0 to 1 scores, that is, from the lowest to the highest levels of government transparency (WJP OGI, 2020). In this regard, Sweden has demonstrated really impressive results in the constraints on government powers (0.91), absence of corruption (0.90), open government (0.86), fundamental right (0.87), order and security (0.91), regulatory enforcement (0.84), civic justice (0.84) and criminal justice (0.87), resulting in an average score of 0.84, which is the highest in the world so far, automatically declaring the country as a global leader in advancing the rule of law.

7 The Content Analysis of Independent Open Data-Driven Projects in Sweden

7.1 The Trafiklab Project: Promoting Open Innovations in e-Transportation Areas

The promotion of smart transportation solutions in Sweden through the reuse of various open data-driven innovations is what exactly the Trafiklab is aimed at. Developed with the direct help from the regional public transport authorities, this open data-driven platform provides a promising venue to advance various types of datasets and, more importantly, application programming interfaces (API) related to the development of smart public transportation in this Nordic country (The Trafiklab project, 2020). Sharing, collaboration and co-development are what this project is based on. By providing tools and information such as open source codes, which independent developers may need to accelerate the initiation and development of a wide range of data-driven e-transportation startups and innovative software solutions in the area, this platform demonstrates an interesting example of how a centralized approach could be harnessed by local transportation companies and regulators in the public transportation industry to promote open data-driven innovations. Indeed, by releasing a wide range of public datasets and, more importantly, by pre-processing the e-transportation datasets so that they could be rendered machine readable and thus easily used in an automatic mode by open data innovators, journalists, academia and technically savvy citizens in developing their own third-party projects, the Trafiklab project is an effective incubator of public sector innovations in the area.

7.2 The Lagen Project: Promoting the Principles of Open Legislation

The Lagen project is an open data-driven initiative that automatically gathers datasets from various government sources and presents all data about the current legislation in a way that is easily comprehensible by citizenry (Lagen, 2020). The platform welcomes contributions from lawyers and citizens on how to make the project better and more comprehensible, actively resorting to the promising power of civic engagement and citizen-sourcing as a tool of quality management by using wiki-based technologies, peer-to-peer networking and encouraging people to reuse the data from the platform in other third-party open data-driven applications and civic engagement projects. As an aggregation tool that collects government data and sorts it in accordance with the sphere, the Lagen platform offers an interesting algorithm of how to promote open data-driven strategies by acting as a certain intermediary between government agencies and developers. In this regard, it is necessary to note that it is not just a duplication of official open data portals but more an aggregation and presentation of all datasets on legislation from one venue and in a more structured way.

Again, as with the Finnish case, the trend demonstrates that open data could create a very interesting model for the prospective promotion of new business niches in the market, potentially leading to a universal self-sustainable mechanism of its realization or even its development as a new platform to advance public sector innovations in the future. In this regard, both developers and business entities could be equally interested in the development of such mechanisms since the former could be more motivated in reusing open data, which is usually published initially by government agencies, creating new commercial products and services based on processed raw datasets and offering them in a more structured and easily comprehensible formats to businesses and other developers, and the latter could create new business or e-commerce solutions in the market. Other developers could also purchase such products for their own services, resulting in the development of more sustainable ecosystems for the realization of the concept and more expedited generation of various open data-driven projects and startups. In this respect, the ultimate direct beneficiaries are both government agencies and civil society, which could enjoy an increase in related public sector innovations, as a result of a growing number of independently developed digital products in the market (see Fig. 5.6).

7.3 *The Open Aid Project: Increase the Transparency of the Public Humanitarian Activity*

Sweden is universally well-known as an active supporter for various development programs to help many developing countries in various relief and

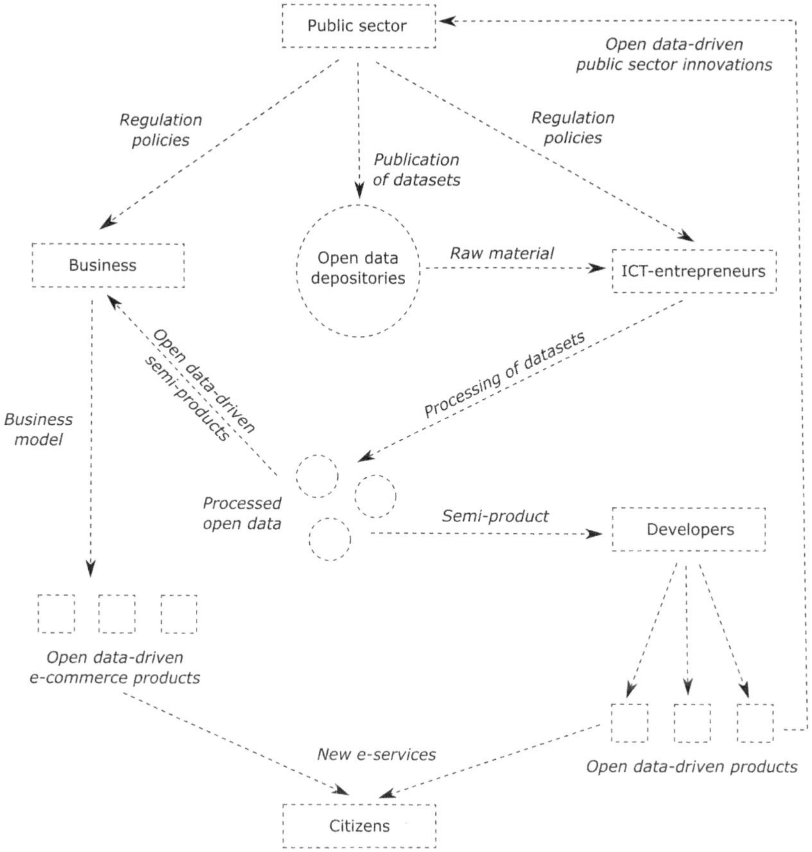

Fig. 5.6 The potential evolution of processed open data as a semi-product service. (Source: Own illustration)

humanitarian programs. In this regard, in order to ensure better transparency of an extremely wide range of international aid programs administered by the central government of the country, public officials launched a special open data-driven project named OpenAid. The platform aims to publish all datasets related to the financial activity of various humanitarian agencies within these international programs in an intrinsically centralized manner (Open Aid, 2020). One of the interesting aspects of the platform is that it helps to sort the activities and programs in accordance with the organization, type of humanitarian aid and countries where it is distributed in a clear visualization form, using mapping and various charts for easy comprehension of data. Hosting thousands of visualizations on various humanitarian activities of Swedish government, this open data-driven project provides a very cost-effective and sustainable way to raise public awareness about all operations of government in the area, using the potential of open data as a raw material for other interactive platforms and information products which could be created by independent developers, data journalists or non-governmental organization for their purposes in the sphere. For example, in 2020 alone, Sweden spent more than US $6 billion for various humanitarian aid missions and relief programs around the world, especially in countries such as Afghanistan, Somalia, Mozambique, Congo and Uganda, among others. The Ministry for Foreign Affairs, The Swedish International Development Cooperation Agency, The Folke Bernadotte Academy, The Swedish Institute and The Swedish Research Council are among the most active public stakeholders in the area.

8 Discussion: Challenges of the Open Data-Driven Movement and Recommendations

8.1 Unsustainable Mechanisms of Funding in the Private Sector

One of the primary challenges that open data-driven projects face today in Sweden, especially independent ones, is a lack of sustainable funding to keep the maintenance of many digital platforms whether they are traditional portals, hybrid web-applications or mobile apps. The current models of the concept promotion are simply non-sustainable financially, even despite the fact that this country provides quite a favorable socioeconomic context to advance in various dimensions and people are morally ready to contribute with their creativity, time and skills to the development of various projects in the sphere. But projects require continuous and long-term support and technical maintenance. In this regard, many promising civic projects again and again repeat the same path of classic failures—when they are successfully launched, often with related extensive publicity in media and social networks, being kept operative for some time and then abandoned or closed, creating multiple dead links as a certain reminder of the unfulfilled promises in the area. Funding is usually a main issue here.

It is a global trend since independent projects are initiated and created by citizens themselves and thus as a voluntary process where self-funding is often the only way to go, such projects repeat the same fate as many innovative startups. In this regard, an often praised and propagated public value of the open data concept, that is, voluntarism, which helps the public sector save some budget funds due to the promise of citizen-sourcing, crowd-funding, increased participation and growth of civic innovations in the ICT-driven public administration, is still voluntarism, even despite the fact that it could be promoted in a favorable socioeconomic environment of egalitarian settings.

The vast majority of startups all over the world are being abandoned due to the lack of support and basic operation maintenance. For instance, such abandoned initiatives as Skuggrix (parliamentary explorer project) and Lokalpolitik (local government explorer project), which were aimed at promoting the principles of digital democracy and transparency at local levels in Sweden, could be regarded as one of the striking examples of such a notorious trend in the area (Lokalpolitik, 2013; Skuggrix, 2010). The use of professional skills in computer technology and design of technically savvy citizens, often students of various technological universities and schools, who usually represent a huge army of the open data movement, especially at local levels, is a very promising phenomenon that often accompanies the launch of various independent projects in the area, demonstrating the potential power of open data to generate civic engagement and participation. But, as the real practice shows, it is not a sustainable way of concept promotion in a long-term perspective. The lack of adequate funding in many civic peer-to-peer projects is one of the main reasons that could explain often the notorious unsustainability of open data philosophy, especially at local levels. Both public and private support is usually temporary and grant money is often pumped only to the handful of projects in the area, sometimes commercial startups. Smaller civic projects usually have little chances to win app contests or hackathons funding. In this regard, the projects that require less attention and maintenance from the technical point of view, for example, due to the application of better interface layouts and more advanced technological platforms, ideally, with an automotive upgrade and flow of datasets, have better chances to survive the vicious circle that could be tentatively called *a technological half-life of open data-driven projects*.

Policy recommendations

(a) *Promotion of more sophisticated funding options from public, private and non-government organizations for open standardized software solutions in the area*

The solution might be found in the development of more sophisticated mechanisms of public and private funding in the area, since the app contests and competitions do not always guarantee the successful continuation of open data-driven projects, especially smaller ones that could not rely on commercial revenue or any other support such as from advertisement and product placements. In this regard, app contests are more suitable for the promotion of young startups that mostly play a role of the venue for professional networking rather than a source of systematic funding and support in a long-term perspective. Public relations and, more importantly, the development of the universal unification of software platforms for the data management and maintenance should play a more important role in this respect, especially when it comes to the publication of source codes for these solutions. Naturally, they should be based on open source code and standardization should become a normal practice in the open data-driven market. For example, the role of non-governmental organizations, businesses, charity funds, venture funds and even the public sector should become more important for the successful realization of the open data concept as a sustainable socioeconomic phenomenon. Direct public support of initiatives that promote more standardized software solutions in the area, preferably based on open source, should be accepted by all stakeholders, especially when such projects do not aim at generating commercial value.

(b) *Donation of abandoned projects, their open source codes and associated databases to open data consortiums*

Collaboration and peer-to-peer support could be one more solution for the challenges with abandoned or discontinued open data-driven projects. In order to understand the collaborative potential of mutually beneficial networking in the area, one should imagine an

(*continued*)

(continued)

open data ecosystem with two possible scenarios for the concept implementation, one with an unsustainable model of realization where only the sense of voluntarism and enthusiasm among data developers that drive the whole movement, which eventually leads to the discontinuation of many open data-driven projects due to no funding and motivation as it was described in two of the cases in Sweden. In this regard, another more optimistic scenario is when any discontinued open data-driven projects are revived and actively supported by other enthusiasts and developers in the area due to the promotion of better funding options. It is possible to create a special consortium of independent developers who could monitor any potentially failing projects and startups in the industry and, if needed, ensure the proper transfer of the source codes of these abandoned or discontinued civic initiatives to the public domain for anyone to adopt and continue them. Such an independent consortium of professional developers could be supported by charities, non-governmental organizations, scientific grants, public fundraising activities or even direct investments from public and private sectors. Such transition could even imply the change of the copyright ownership (though the author's rights could still remain intact even after the transfer of the copyright and author's names could always be accredited and respected) on either philanthropic or commercial basis due to the fact that the initial developers of open data-driven projects, who might not be able to continue anymore their normal operation due to some objective or subjective reasons, may wish to donate or sell the results of their development for a symbolic price to non-governmental players in the market. Moreover, the original developers could hand over the original source code of such applications or platforms to an open community of developers under the conditions of franchise or even common license such as under the Creative Commons copyright rule, as in case with free distribution of open source software solutions in Linux operating systems, so that not to interrupt the successful life cycle of open data-driven public sector innovations and sustainable generation of new digital services, startups and information products, which could be valuable for political or socioeconomic development of local communities (see Fig. 5.7).

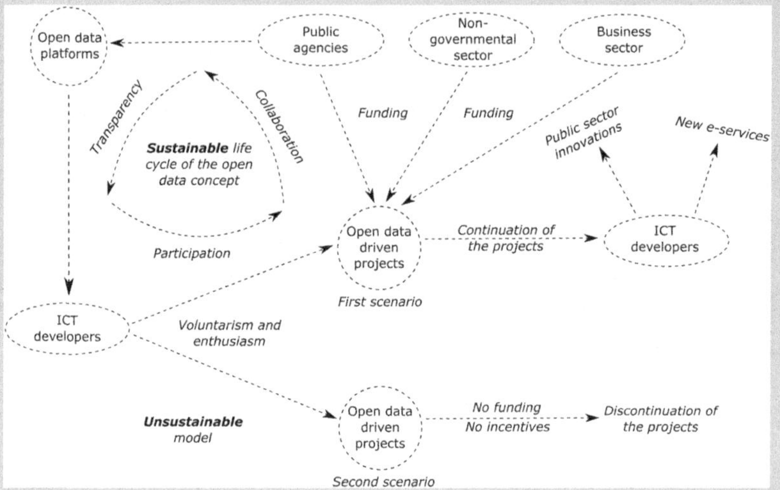

Fig. 5.7 The sustainable and unsustainable mechanisms in the promotion of open data as a socioeconomic phenomenon. (Source: Own illustration)

8.2 Outdated Information and Quality of Datasets

Another challenge is that the open data-driven projects often contain outdated information, especially if their creators have no time or financial opportunity to keep their platforms updated and maintained in an interactive mode in accordance with the latest technological developments in the area. It is necessary to note that pre-processing of open data could be quite a time- and money-consuming undertaking. Because of lack of expertise and professional software to deal with certain formats of data, over time many projects become static and obsolete when the information and datasets become older. The automatic update and synchronization of datasets, which could aggregate them in an interactive mode from public depositories and platforms, might fail due to the changed web addresses of public institutions or broken links that, as it appears, need to be checked on a regular basis. The datasets published by the government are not always easily machine readable and often require further processing, which in turn creates additional technological barriers for independent developers, resulting in higher costs in operating various open data-driven projects and platforms. It is especially true when the information and datasets are derived from the archives such as data with visualization, old statistical charts, old photos and topographic maps, handwritten registries and documents, audiovisual files in old formats and so on. The reuse of these types of information often require not only special software, often proprietary, closed and, therefore, expensive one, but also special skills and expertise from the part of developers and usually is a very time-consuming undertaking, which could make the pre-processing of semi-datasets extremely costly.

Policy recommendation

(a) *Automatic publication of FOIA requests and creation of standardized APIs for the interactive update of government datasets in public depositories*

The freedom of information policy provides a powerful instrument that allows citizens to request a wide range of datasets from various public agencies. However, such data is usually released individually to each of the requests and exactly to a person who requested it. In this regard, it would be more effective if the principle of real-time automation was used to update public open data platforms in an intrinsically centralized manner, once these requests are addressed by corresponding government agencies so that to open the content of FOIA datasets and avoid multiple duplications of requests sent from citizens that are related to the same type of released information. Naturally, such publication of open data should be allowed only if these datasets do not contain any personal information, commercial and state secrets or any classified information, for example, related to national security or defense.

(b) *Creation of the special pool of professional software products for automated selection and pre-processing of various raw datasets*

Another possible answer to these challenges might be found in developing a special pool of professional software products for automated selection and, if necessary, processing of various raw datasets

(*continued*)

(continued)
so that they could be easily updated in a machine-readable way. Naturally, the access to this pool should be allowed only to the members of the special consortium of open data developers. The wide community of independent developers all over the world could provide a promising venue to seek necessary expertise and playground to test various solutions in the area. The distribution of such digital instruments should be on a free-of-charge basis and the generation of new digital services and platforms should be regarded as a main public value of such cooperation among developers, designers, coders and the like on a global scale. For inspiration one should look at such universally well-known depositories of open source software as Sourceforge, Bitbucket or GitHub, which allow members of their communities to create, test and distribute in a transparent and collaborative manner various digital platforms and software solutions. There should be a global networking venue created specifically to meet the needs of open data developers.

9 Summary: Key Findings

This chapter of the book was, first of all, aimed to explore a truly multidimensional nature of open data from the context of a country that is globally well-known today for its exceptionally liberal policies in the area of freedom of information, especially in providing almost unlimited and unrestricted public access to various types of government information and datasets. In this regard, Sweden is universally regarded today as a global leader of the open data movement, which actual success in the area as the case study demonstrates could be explained not only by its extended open data policies that are actively promoted and propagated in almost all public sector reforms but also by a combination of favorable ideological, political and technological preconditions that provided an ideal socioeconomic playground to advance the concept further in the context of social democracy. The development of policies aimed at distributing wealth more evenly across different layers and strata of population has provided a promising socioeconomic environment to network various stakeholders around the open data movement in this country and accelerate related public sector innovations, for example, by resorting to the power of civic engagement and altruistic sentiments of members of civil society that are extremely strong today in this egalitarian society due to the widespread philosophy of social equality and humanism that is shared by many citizens. The socioeconomic readiness of open data enthusiasts, independent developers and technically savvy citizens, who are driven mostly by a widespread sense of community and voluntarism, to join and lead the movement by helping their local neighborhoods and communities to thrive and progress has played a crucial role in the popularization of the concept in this Nordic nation. By networking with each other and other stakeholders from public and private sectors as well as with representatives from non-governmental institutions and think tanks, they often try to experiment with various platforms and startups, seeking the most sustainable models of the concept development.

References

Ackerman, J. M., & Sandoval-Ballesteros, I. E. (2006). The global explosion of freedom of information laws. *Administrative Law Review, 58*, 85.

Andersson, P., Hulten, S., & Valiente, P. (2005). Beauty contest licensing lessons from the 3G process in Sweden. *Telecommunications Policy, 29*(8), 577–593.

Appelgren, E., & Nygren, G. (2014). Data Journalism in Sweden: Introducing new methods and genres of journalism into "old" organizations. *Digital Journalism, 2*(3), 394–405.

Axelsson, A. S., & Schroeder, R. (2009). Making it open and keeping it safe: E-enabled data sharing in Sweden. *Acta sociologica, 52*(3), 213–226.

Berggren, C., & Laestadius, S. (2003). Co-development and composite clusters – The secular strength of Nordic telecommunications. *Industrial and Corporate Change, 12*(1), 91–114.

Blomstrom, M., & Kokko, A. (2003). From natural resources to high-tech production: The evolution of industrial competitiveness in Finland and Sweden. The World Bank. http://web.worldbank.org/archive/website00895A/WEB/PDF/BLOMSTRO.PDF

Burkart, P. (2014). *Pirate politics: The new information policy contests.* MIT Press.

Chaminade, C., & Edquist, C. (2006). *Rationales for public policy intervention from a systems of innovation approach: The case of VINNOVA.* Lund University.

Cool, A. (2015). *Detaching data from the state: Biobanking and building Big Data in Sweden.* BioSocieties.

CTSW. (2020). The Civic Tech Sweden workshops. https://civictech.se/en/events/

Dopter. (2020). http://dopter.se/

Dvorak, J. C. (2006). The politics of piracy emerge in Sweden. PC Magazine Online.

Elder, N. C. (2013). *Government in Sweden: The executive at work*. Elsevier.
Eriksson, K., Ernkvist, M., Laurell, C., Moodysson, J., Nykvist, R., & Sandström, C. (2019). A revised perspective on innovation policy for renewal of mature economies – Historical evidence from finance and telecommunications in Sweden 1980–1990. *Technological Forecasting and Social Change, 147*, 152–162.
Erlingsson, G. Ó., & Persson, M. (2011). The Swedish pirate party and the 2009 European Parliament election: Protest or issue voting? *Politics, 31*(3), 121–128.
ESH. (2020). The East Sweden Hack. http://eastswedenhack.se
Fink, K. (2018). Opening the government's black boxes: Freedom of information and algorithmic accountability. *Information, Communication & Society, 21*(10), 1453–1471.
Global Open Data Index. (2019). http://index.okfn.org/
Grundstrom, C., & Lövnord, A. (2014). Open Data: Barriers when adopting, implementing, and innovating open data in local municipal organizations. Master thesis, The Umea University. SPM 2014.19. http://www.diva-portal.org/smash/record.jsf?pid=diva2%3A726112&dswid=-3671
Hack the Crisis. (2020). The International App contest to fight the consequences of the coronavirus crisis in economy and society through innovations. https://www.ai.se/en/events/hack-crisis-0
Handlingar Project. (2020). The FOIA requests platform. https://handlingar.se/
HAP. (2020). The Hackaway project. http://www.hackaway.com/
Heckscher, E. F., & Heckscher, G. (1954). *An economic history of Sweden*. Harvard University Press.
Henrekson, M., & Jakobsson, U. (2005). The Swedish model of corporate ownership and control in transition. http://papers.ssrn.com/sol3/papers.cfm?abstract_id=998256
HFS. (2020). The Hack for Sweden. http://hackforsweden.se/
HHD. (2015). The Health Hack Day. http://healthhackday.com/
Hultkrantz, L. (2002). Telecommunications liberalisation in Sweden: Is "intermediate" regulation viable? *Swedish Economic Policy Review, 9*, 133–161.
Jääsaari, J., & Hildén, J. (2015). Piracy & social change| from file sharing to free culture: The evolving agenda of European pirate parties. *International Journal of Communication, 9*, 20.
Jacob, M. (2006). Utilization of social science knowledge in science policy: Systems of innovation, Triple Helix and VINNOVA. *Social Science Information, 45*(3), 431–462.
Janssen, K. (2012). Open government data and the right to information: Opportunities and obstacles. *The Journal of Community Informatics, 8*(2).
Joshi, D., & Navlakha, N. (2010). Social democracy in Sweden. *Economic and Political Weekly, 45*, 73–80.
JPP. (2019). The Journalism++ Project. http://jplusplus.se/

Juell-Skielse, G., Hjalmarsson, A., Johannesson, P., & Rudmark, D. (2014). Is the public motivated to engage in open data innovation? In *International conference on electronic government* (pp. 277–288). Springer.

Kassen, M. (2015). e-Federalism and e-Centralism: Key findings and recommendations. In *Understanding systems of e-government* (pp. 132–155). New York, NY.

Khayyat, N. T. (2017). A study of telecommunication policies and broadband penetration for Sweden and South Korea. *UKH Journal of Science and Engineering, 1*(1), 26–38.

KHD. (2020). The Kids Hack Day. http://www.kidshackday.com/

Klein, T., Langner, J., Broman, B., Svensson, J., Frankenberg, B., Langborg, T., & Bennet, C. (2012, April). Environment Climate Data Sweden-Facilitating a wider sharing of environment and climate data. In *EGU general assembly conference abstracts* (Vol. 14, pp. 59–91).

KSC. (2020). The Kista Science City. www.kista.com

Lagen. (2020). The open legislation project. www.lagen.nu

Larsson, S., & Malmberg, A. (1999). Innovations, competitiveness and local embeddedness: A study of machinery producers in Sweden. *Geografiska Annaler: Series B, Human Geography, 81*(1), 1–18.

Lassinantti, J. (2014). Relevant social groups within the open data initiative – A matter of intertwining meanings. https://pure.ltu.se/portal/files/65258656/iris36_submission_47.pdf

Lassinantti, J., Bergvall-Kåreborn, B., & Ståhlbröst, A. (2014). Shaping local open data initiatives: Politics and implications. *Journal of Theoretical and Applied Electronic Commerce Research, 9*(2), 17–33.

Lfstedt, U. (2007). E-Government services in local governments a study of development in Swedish municipalities. *Journal of Organisational Transformation & Social Change, 4*(2), 157–176.

Li, M. (2009). Pirate party and the pirate bay: How the pirate bay influences Sweden and international copyright relations. *Pace International Law Review Pace International Law Review, 21*, 281.

Linders, D. (2013). Towards open development: Leveraging open data to improve the planning and coordination of international aid. *Government Information Quarterly, 30*(4), 426–434.

Lindgren, S., & Lundström, R. (2011). Pirate culture and hacktivist mobilization: The cultural and social protocols of# WikiLeaks on Twitter. *New Media & Society, 13*(6), 999–1018.

Lokalpolitik. (2013). An example of the closed open data-driven project. http://www.lokalpolitik.se/

LU. (2020). Linköping University. http://www.liu.se/?l=en

LUII. (2020). Lund University Internet Institute. http://luii.lu.se/research/legal-solutions-for-open-data

Lundqvist, M. A., & Williams Middleton, K. L. (2010). Promises of societal entrepreneurship: Sweden and beyond. *Journal of Enterprising Communities: People and Places in the Global Economy, 4*(1), 24–36.

Lundström, T. (1996). The state and voluntary social work in Sweden. *Voluntas: International Journal of Voluntary and Nonprofit Organizations, 7*(2), 123–146.

Mecherkany, R. (2010). *Social entrepreneurship in Sweden*. Royal Institute of Technology.

Miegel, F., & Olsson, T. (2008). From pirates to politicians: The story of the Swedish file sharers who became a political party. In *Democracy, Journalism and Technology: New Developments in an Enlarged Europe* (pp. 203–216). Tartu University Press.

MSP. (2017). The Meta Solutions Project. http://metasolutions.se/tag/project/

Mustonen, J. (2006). *The world's first freedom of information act: Anders Chydenius' legacy today*. Anders Chydenius Foundation.

Nationalmuseet. (2021). The Swedish National Museum. https://www.nationalmuseum.se/en/samlingarna/fria-bilder

NITH. (2020). The Nordic Internet of Things Hackathon. http://nordiciothackathon.com/

Nordenstreng, K. (2017). Freedom of speech in Finland 1766–2016: A Byproduct of Political Struggles. The Legacy of Peter Forsskål: 250 Years of Freedom of Expression.

Nordin, J. (2016). The Swedish freedom of print act of 1776 – Background and significance. *Journal of International Media & Entertainment Law, 7*, 137.

Obama, B. (2009). Memorandum for the heads of executive departments and agencies. *Presidential Studies Quarterly, 39*(3), 429.

ODMS. (2020). The open data meetups in Stockholm. http://open-data.meetup.com/cities/se/stockholm/

OECD. (2019). Government at a glance 2019: Sweden. http://www.oecd.org/sweden/govataglance.htm

OGP. (2015). The open government partnership. The commitments of Sweden. http://www.opengovpartnership.org/country/sweden

Olafsson, G. (2012). Call for open humanitarian information. Guest Editors, 98.

Open Aid. (2020). The OpenAid open data-driven project. http://openaid.se

Open Lund. (2020). The official open data project of the city of Lund. http://www.lund.se/psidata

Open Malmö. (2020). The open data platform of the city of Malmö. http://se-city.census.okfn.org/place/malm%C3%B6

Open Stockholm. (2020). The official open data project of Stockholm. http://dataportalen.stockholm.se/dataportalen/

Open Umeå. (2020). The open data platform of the city of Umeå. https://opendata.umea.se/pages/startsida/

Open Uppsala. (2020). The official open data project of the city of Uppsala. https://www.uppsala.se/psidata

Oppnadata. (2020). The official open data project of Sweden. https://www.dataportal.se

PDP. (2017). The Popdevelop project. The networking platform of ICT developers from Malmo, Sweden. http://popdevelop.com/

RIA. (2020). The Research Institute Acreo. https://www.acreo.se

Rudmark, D., & Hjalmarsson-Jordanius, A. (2019). *Harnessing digital ecosystems through open data-diagnosing the Swedish public transport industry*. Stockholm & Uppsala, Sweden, European Conference of Information Systems (ECIS), Association for Information Systems.

Ruuska, I., & Teigland, R. (2009). Ensuring project success through collective competence and creative conflict in public–private partnerships – A case study of Bygga Villa, a Swedish triple helix e-government initiative. *International Journal of Project Management, 27*(4), 323–334.

Rydholm, L. (2013). China and the world's first freedom of information act: The Swedish freedom of the Press Act of 1766. *Javnost – The Public, 20*(4), 45–63.

Safarov, I. (2019). Institutional dimensions of open government data implementation: Evidence from the Netherlands, Sweden, and the UK. *Public Performance & Management Review, 42*(2), 305–328.

SGS. (2020). The Startup Grind Stockholm. http://www.meetup.com/Startup-Grind-Stockholm/

SICS. (2020). The Swedish Institute of Computer Science. https://www.sics.se/about-sics

Skuggrix. (2010). The Skuggrix parliamentary transparency and civic engagement project. https://old.liu.se/liu-nytt/arkiv/mn/1.179467?l=sv

SS. (2020). The Swedish Software. http://www.swedishsoftware.se

SSS. (2017). The Swedish Startup Space. http://swedishstartupspace.com

Steinmetz, K. F. (2012). Wikileaks and realpolitik. *Journal of Theoretical & Philosophical Criminology, 4*(1), 14.

Stråle, J., & Lindén, H. (2014). *An evaluation of platforms for open government data. Degree project in computer engineering*. Royal Institute of Technology.

Susha, I., Grönlund, Å., & Janssen, M. (2015). Organizational measures to stimulate user engagement with open data. *Transforming Government: People, Process and Policy, 9*(2), 181–206.

Swedsoft. (2020). The Swedish Innovation Center. http://swedsoft.se/

Tele2 87. (2017). The Tele2 Hack Day. https://tele2.confetti.events/

The Swedish National Data Service. (2020). https://snd.gu.se/en

The Trafiklab Project. (2020). The open data-driven e-transportation project. https://www.trafiklab.se/

Timpka, T., Spreco, A., Dahlström, Ö., Eriksson, O., Gursky, E., Ekberg, J., ... Holm, E. (2014). Performance of eHealth data sources in local influenza

surveillance: A 5-year open cohort study. *Journal of Medical Internet Research, 16*(4), e116.
TMS. (2020). The TechMeetups Stockholm. http://www.meetup.com/Stockholm-TechMeetups/
Trägårdh, L. (2007). *State and civil society in Northern Europe: The Swedish model reconsidered* (Vol. 3). Berghahn books.
UN. (2020). United Nations E-Government Survey 2020. https://www.un.org/development/desa/publications/publication/2020-united-nations-e-government-survey
Vamstad, J., & von Essen, J. (2012). Charitable giving in a universal welfare state – Charity and social rights in Sweden. *Nonprofit and Voluntary Sector Quarterly*. 0899764012466176.
Vinnova. (2020). Sweden's innovation agency. http://www.vinnova.se/en/
Wallmark, J. T., & McQueen, D. H. (1991). One hundred major Swedish technical innovations, from 1945 to 1980. *Research Policy, 20*(4), 325–344.
Weaver, C., Davenport, S., Baker, J., Findley, M., Peratsakis, C., & Powell, J. (2014). *Malawi's open aid map*. World Bank.
Widholm, A., & Appelgren, E. (2020). A softer kind of hard news? Data journalism and the digital renewal of public service news in Sweden. New Media & Society, 1461444820975411.
WJP OGI. (2020). World justice project – Open government index. https://worldjusticeproject.org/rule-of-law-index/country/2020

CHAPTER 6

Conclusion: Understanding the Collaborative Nature of Open Data Governance

Abstract The key goal of the chapter is to analyze and elaborate further on key roles of various stakeholders, including public agents, civic communities, business and corporate players, professional developers, nongovernmental sector, academia and journalistic circles. As it was demonstrated by both Finnish and Swedish cases, all these actors may have different interests in the promotion of open data governance as a collaborative movement. As such, their contributions and contributing roles differ in a number of important aspects. These nuances of the collaboration and networking are discussed in more detail in this part of the research. Toward the end of the chapter, the author concludes on key theoretical and practical contributions of the research to current academic knowledge on the topic and discusses limitations and potential future agendas for research in this direction.

Keywords Open data • Conclusion • Stakeholders • Collaboration • Networking

M. Kassen, *Open Data Governance and Its Actors*, Studies in National Governance and Emerging Technologies,
https://doi.org/10.1007/978-3-030-92065-4_6

1 Understanding the Roles of Various Stakeholders in the Development of Open Data

1.1 Research Question # 1: Understanding the Decisive Role of the Public Sector

By initiating various open data government projects and providing various types of datasets in a transparent manner via special freely accessible repositories, the public sector presumably sends a strong political signal for other stakeholders to join the movement and motivates them to reuse open data. In this regard, it seems interesting to understand how the government acts as a supernode in uniting the interests of various actors and creating new promising political networks between them.

Although open data is aimed to advance civic engagement and greater autonomy of civil society and businesses in developing their own data-driven information products and services for peers, including in e-government related areas, the public sector still plays a crucial role of not only as a dataset provider by publishing open data on special repositories but also as a initiator of various public awareness and public relations campaigns such as hackathons, prize contests and development competitions. These initiatives are all aimed to propagate useful information among potential users of open data, especially computer specialists and developers (e.g. computer programmers, designers, coders, system administrators, students of information, library and computer sciences as well as technically savvy citizens who often call themselves independent developers). Often these events are organized in close cooperation with the non-governmental organizations, private sector, academic and journalistic communities and all other stakeholders.

Moreover, various centralized development platforms are directly created and supported by the public sector to network various stakeholders around the open data movement and other e-government related areas. For example, as it was demonstrated in the case of the Forum Virium (Finland) and Kista City initiatives (Sweden), by providing information, financial and organizational support for various startups in the area, the government in both Nordic countries truly sends a strong signal that all prospective and promising projects will be supported in a centralized manner through these collaborative and networking platforms. This is especially important since decision-making is often promoted in a centralized way in both contexts, especially in Finland, given the unitary nature of government institutions there. In this regard, both national and local government bodies (e.g. ministries, national agencies, city halls and municipalities) participate in the advancement of various open data development programs across the nations and across various institutions with an ultimate goal to generate raw datasets, which could be then published through various public data repositories at both national and local levels and ultimately reused by other stakeholders (see Fig. 6.1).

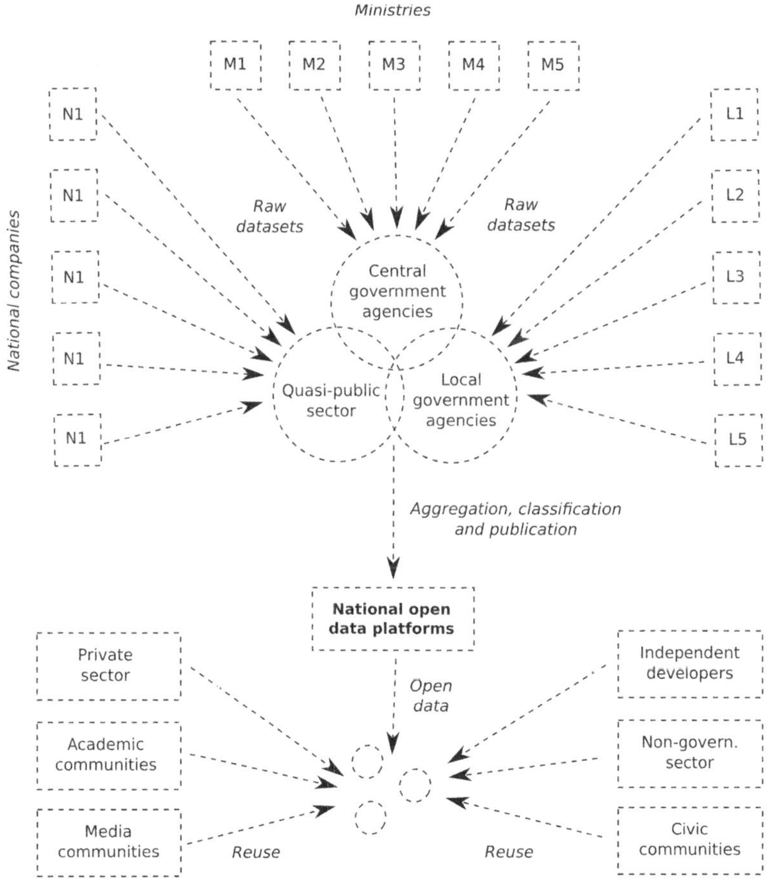

Fig. 6.1 Understanding the multi-institutional role of public sector in advancing open data governance. *Source: own illustration*

One of the interesting trends that could be observed in two country cases is that government agencies are tentatively beginning to develop various public feedback platforms on these data repository platforms to identify the need for certain types of datasets for both civic and corporate stakeholders. Three motivations drive the open data governance movement among public agencies: (a) further democratization of government information processes through the promotion of greater transparency in publishing public information and revealing all details associated with decision-making at all institutional levels of power; (b) the promotion of public sector innovations through the initiation of various civic engagement and e-participation projects which all could be driven by open data, though mostly in urban areas; and (c) privatization of a number of public services which could be now developed and promoted by civic and corporate players through new open data-driven products and platforms, which could be independently developed without the direct participation of government in the process due to the emancipating power of open data governance.

1.2 Research Question # 2: Understanding the Contributing Role of the Civil Society

Taking into account the highly collaborative nature of open data governance, it seems interesting to understand how such new data-driven networking layers as citizens-to-citizens (e.g. independent peer-to-peer open data projects), citizens-to-businesses (e.g. open data-driven e-commerce projects) and, more importantly, citizens-to-NGOs (e.g. the participation of people in various non-governmental initiatives, including those that operate in the open data area), citizens-to-media (e.g. the participation of citizens in the public scrutiny of government agencies through mass media) and citizens-to-developers (e.g. the cooperation of people with developers in launching various open data-driven independent projects, sharing and crowd-sourcing platforms) are manifested in real-life contexts.

Civil society plays an important role in the advance of the open data concept as, first of all, an ideological platform to promote digital democracy and greater transparency of information processes in government. Open knowledge is a key conceptual paradigm that shapes the whole movement in both countries, which could be manifested in the popularity of such public initiatives as open science, open society, open state, open economy, open government and open journalism. In this regard, open data is widely regarded in both countries as one of the most important directions of the open knowledge movement, which is aimed at propagating the public values of the open government concept and, to a certain degree, is considered as an instrument to achieve greater transparency in government. In this regard, both Finnish and Swedish cases clearly demonstrated that open data is a highly context-dependent phenomenon, which requires the readiness of both public and civic sectors to adopt the concept as one of the prospective ways to increase greater communication between government and civic society through various networking events that could be organized in the area and, more importantly, through the collaboration and cooperation in the development of open data-driven projects and initiatives.

As a context-dependent phenomenon, open data requires greater participation from civic actors on a voluntary basis. For example, these independent developers have to invest their time, skills, knowledge and often significant financial resources to successfully run data-driven platforms and initiatives. Understandably, more egalitarian societies such as Sweden, Finland and other Nordic countries have greater conditions and prospects to equip its citizens with opportunities to participate in the open data movement as not only consumers of the data-driven services and information products but also providers of additional data (civic data) and ultimately contributors and developers themselves. Social democracy and centuries-old traditions of government transparency aimed to advance the values of equity, social justice, compassion and adherence to a society that protects all its members create favorable political and socioeconomic contexts to promote open data governance as a highly collaborative phenomenon that networks various interested stakeholders around one important idea to make governance better and more effective in achieving its goals and values. These values are traditionally shared and supported by the majority of the population in egalitarian societies and reflected in the decisions made by its representatives in government.

1.3 Research Question # 3: Understanding the Active Role of Business Communities

Open data governance could hypothetically reduce the presence of the public sector in the e-service procurement market, leaving more space for the private sector to expand and prosper due to small and more effective technology-driven government and related public service ecosystems it has to operate in.

In this respect, it is interesting to test these assumptions and understand what the primary motivations of the business actors are to join the open data movement and advance the concept as a collaborative and networking phenomenon, especially in cooperating and partnering with various stakeholders, both public and civic.

The business and corporate sectors play important roles as both an intermediary and a promoter of various open data-driven initiatives. As such, as it was demonstrated in both country cases the motivations and contribution of private stakeholders could also be diverse and multidimensional, ranging from the mere intention to commercialize the potential of open data in various sectors of economy and ending to the need to raise public awareness about the phenomena and fund-related prospective projects which were initially proposed by civic and non-governmental sectors. Such interesting concepts as an algorithm-as-a-service (data-driven open software and open innovations) and the pre-processing of raw datasets for other stakeholders clearly demonstrate the potential of the open data movement to accelerate business processes in specific niches of the information technology market. Paradoxically, by developing these products and driven by commercial interests and motivations, private and corporate developers not only attract the interest of the market and consuming society to the open data movement but also indirectly promote the initiation and development of more centralized and, more importantly, standardized solutions in the area, which in turn reduces the cost of operation and maintenance in the area, greatly boosting related innovations and competition in the area, which could eventually benefit other actors in the open data movement (see Fig. 6.2).

The active participation of the private sector in the movement showcases that the open data governance concept is not only related to the reuse and management of published datasets but also to the reuse and reapplication of various new software products, patterns of software codes and algorithms that could be adapted to various goals and needs of clients, including public and civic ones from government and non-governmental sectors of economy, helping to advance healthy competition and knowledge building in such once monopolized and thus highly lopsided areas as e-government and e-commerce. Indeed, by offering more flexible, universal and adjustable software solutions that could utilize various government and civic data, these private stakeholders create a new market niche which is specifically focused on the development of more cost-effective information products and platforms to advance open data as a lucrative business

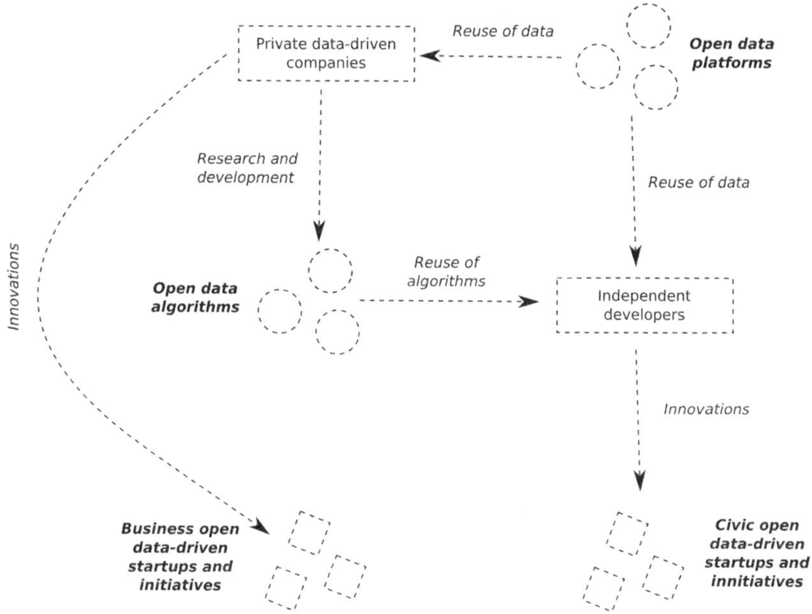

Fig. 6.2 Understanding the role of the private and corporate sectors in promoting open data-driven innovations. *Source: own illustration*

model to follow, sending a strong signal to various stakeholders about the great socioeconomic potential of the concept to innovate current economic models in the software industry. In both Finland and Sweden, the development of the software industry plays a crucial role in achieving national competitiveness in various sectors of the economy due to the high degree of automation in many industries. In this regard, the concept of open data along with the promotion of big and linked data provides great opportunities to optimize and improve current production models and realization strategies.

Both small- and middle-size enterprises and corporate players are beginning to pay attention to the promise of open data to open new economic niches in various sectors of the economy through data-driven innovations. More importantly, given the highly context-dependent and collaborative nature of the concept, influential stakeholders in the area start to directly fund and technologically support various startup projects and civic initiatives that are driven by the reuse of open government datasets and open software solutions. Public-private partnerships in organizing such important networking events as app contests and master classes with the participation of industry leaders are a popular way to support various ventures in the area, including civic ones. After all, the successful development of the commercial market for data-driven products and services would require the army of those who could contribute with additional data, mostly among civic and government communities. Thus businesses along with public stakeholders are beginning to understand that both public administrative and economic environments will eventually benefit from more active roles of citizens in the process, too, and as such start to support all prospective civic initiatives in the area, not necessarily only those that have an immediate economic or commercial potential.

1.4 Research Question # 4: Understanding the Crucial Role of Software Developers

The computer experts and data programmers in the area could presumably regard open data governance as a technological platform, which embraces the best developments in the ICT sector in order to make public administration and political communication with citizens and businesses better and more transparent or they may simply consider their role as a technological bridge between other key players, which, unfortunately, is often forgotten by many policymakers and widely regarded as an insignificant part of the open data concept. Moreover, many developers could also be dependent players who mostly serve the government in the technological reformation of public administration and, therefore, in this case, they should be regarded as part of the public sector itself. Such assumptions could also be tested in a real-life context.

Open data developers, who represent a wide range of computer specialists, e-government operators, software programmers and associated audiences, ranging from professional government and corporate e-commerce experts and ending to independent civic players in the area, are the true proponents, promoters and, more importantly, drivers of the open data movement. It is open data developers that often on their own initiative create and promote a wide range of data-driven products and solutions for various audiences and for various administrative and socioeconomic needs. New public and commercial products are aimed to improve governance (e.g. in advancing various e-participation, e-democracy, e-voting, e-collaboration, e-education, e-healthcare and other public digital platforms) or advance new economic services (e.g. in promoting a wide range of e-utilities, communal e-services, e-transportation and e-procurement projects).

Both Finnish and Swedish cases demonstrated that both professional and independent civic developers could be engaged in developing a wide range of open data-driven projects in various sectors of the economy, both public and private. In this regard, their contributions to the open data movement could be different, too, depending on their roles in the process. Naturally, the role of independent or civic developers, who are often driven by philanthropic or egalitarian motives, could be quite important as they often initiate, develop and advance a wide variety of open data-driven projects and startups (which could be of a less interest for commercial companies and big corporations), providing various services for local communities. Reuse of datasets is a key process of the development that could meet the need for low market services or certain economically non-lucrative yet publicly or politically important niches in such areas as e-governance or e-democracy. Taking into account that there are no copyright restrictions to reuse government datasets, hence it is named open data, these independent civic developers could reprocess and adapt these public datasets to their needs, both personal and public, create new software products, information services or launch innovative commercial or community projects to start their own businesses or help their local neighborhoods with new solutions and solve various issues at local levels.

As a context-dependent phenomenon, the open data movement requires the army of computer experts and specialists in various areas such as software developers, coders, designers and testers to meet the need for the critical mass of users who could potentially join the movement and start reusing datasets for their own third-party projects or startups. In this regard, the intrinsic focus and increased funding for various education programs that have been aimed to prepare the new generation of these experts in local universities in both Finland and Sweden definitely helped to enrich the market with a much-needed labor force in important software and telecommunication industries. These education policies eventually benefited the open data concept, too, since a fraction of these professionals could potentially join the movement. It is interesting to note that many of the independent open data developers who created various data-driven startups and platforms are often students of local tertiary education institutions. Most of them are majoring in computer sciences.

1.5 Research Question # 5: Understanding the Intermediary Role of the Non-governmental Sector

Other promising venues for empirical analysis could be found in researching such interactions as NGOs-to-citizens and NGOs-to-developers transactions, for example, in helping to engage citizens and developers; NGOs-to-business interactions, especially in an attempt to understand the fundraising activity and diversification of funding options for various open data-driven e-commerce projects; and NGOs-to-media, for example, in identifying the invisible mechanism of open data-driven collaboration that could relate, for instance, to the investigation of cooperative media coverage of various civic engagement programs in the area, public awareness programs and supporting research and related curricula in various institutions of tertiary education.

The non-governmental sector, which is represented by a great number of players, starting from various think tanks and development foundations and ending to academic institutions, is actively engaged in advancing a wide range of open data initiatives in both of the countries. For example, various universities participate in a wide range of innovation projects aimed to make governance better at both national and local levels. Moreover, various education institutions themselves often publish open data in public and private depositories and directly fund various data-driven projects in both Finland and Sweden. For example, by participating in the development of their own open data projects and startups, students of computer science and computer programming majors could receive academic credits. Such support by local academic institutions is really crucial to stimulate the reuse of open data by those who could potentially contribute their time, expertise and effort to create new information products, experiment with the development of various e-government and e-commerce services and projects. Academic communities that are actively engaged in related research on open data often participate in various operations of various think tanks, activities of corporate research and development labs and consult government agencies on related topics. In addition, the open data movement is also enriched and accelerated by the growing number of proponents for open science and open software movements among academic communities. The publication of open scientific data in special public depositories is one of the latest trends in the area.

Participation of non-governmental organizations as well as academic communities and students of computer majors in the organization of various master classes, app contests, technological incubators, hackathons and many other collaborative and networking events helps to advance the concept of open data governance as a truly collaborative and networking phenomenon where the intermediary role of the non-governmental sector is all about delivering knowledge, expertise and organization support to various stakeholders, both public and private. For example, the active cooperation and partnership of local universities with various private corporations helps to advance research and development in various sectors of economy, including in the open data areas since many local telecommunication companies and software firms are directly engaged in funding and helping to prepare an army of computer specialists and support various research projects in the field. As such the popularity of the open knowledge movement, which is reflected in the advance of such phenomena as open science, open innovation and open software, directly benefits open data governance and helps academic communities to promote various e-education open data-driven startups and initiatives in local universities (see Fig. 6.3).

Both Finland and Sweden are following the same trend that is popular in many western democracies where the role of the Open Knowledge Foundation plays an important role of the venue that network the interests of various stakeholders, first of all, from the civic and non-governmental sectors of economy on the one hand and government and private actors on the other, for example, public agencies, thinks tanks, journalists, independent developers, civic and business communities around the idea to make governance more effective and responsive through the publication of information, including as raw open data in special public depositories. Crowd-sourcing and, to a certain degree, crowd-funding are important elements of the movement that help to advance the open data concept as a philanthropic collaborative and networking phenomenon. Indeed, independent developers who sometimes call themselves as technically savvy citizens often have to rely on themselves or cooperation and partnership with other civic peers in developing related open data-driven projects. The lack of economic incentives due to small market niches for related commercial products and services in the area helps to advance the open data concept as a certain benchmark of the socioeconomic readiness of the civil society to contribute, though still leaving open the question about the true motives of civic developers to join the movement.

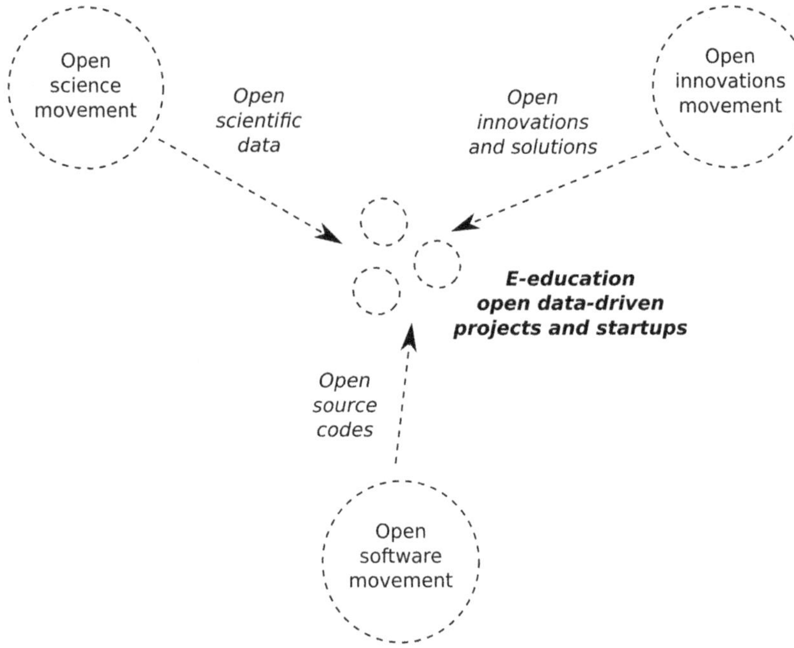

Fig. 6.3 Understanding the public values of the open knowledge movements in academic communities. *Source: own illustration*

1.6 Research Question # 6: Understanding the Information Role of Journalistic Communities

Taking into account the fact that media are beginning to play an ever-increasing role in raising public awareness about open data as a collaborative movement, it is interesting to understand how related journalistic communities use open data in their everyday activities and what motivates them to join the movement and participate in various media coverage campaigns initiated by both public and private actors in the area.

The role of journalistic communities is probably the most overlooked in academic research on open data governance. In this regard, both Finnish and Swedish cases clearly demonstrated that media specialists could play an important role as important intermediaries who directly affect the increase in public awareness about the open data movement in all its diversity. Indeed, a growing number of journalists are beginning to pay close attention to the development of both public and private data-driven projects and initiatives. A wide range of electronic media outlets, television channels and information agencies are starting to create media content that covers topics related to open data phenomena in a variety of sectors of the economy. One of the important roles that media is beginning to play is the provision of official information sponsorship for various professional meetings and forums organized in the area such as app contests, information promotion of innovation clusters, new startups and hackathons. These measures eventually help to raise public awareness about the open data movement and attract the attention of potential investors, new developers and, more importantly, citizens who could directly contribute with their ideas, knowledge and new civic data. In general, data journalism is based on three pillars of related activities in the area, namely the publication of open journalistic data, reuse of open government, corporate and civic data and, finally, the media coverage of open data-driven activities by other stakeholders (see Fig. 6.4).

Indeed, media conglomerates and national companies in the area generate a large volume of their own datasets which are often published by them in their own repositories. The goal is to initiate public sector innovations in the media area, too. In this regard, big data and open data-driven journalism or, for short, data journalism has become one of the modern

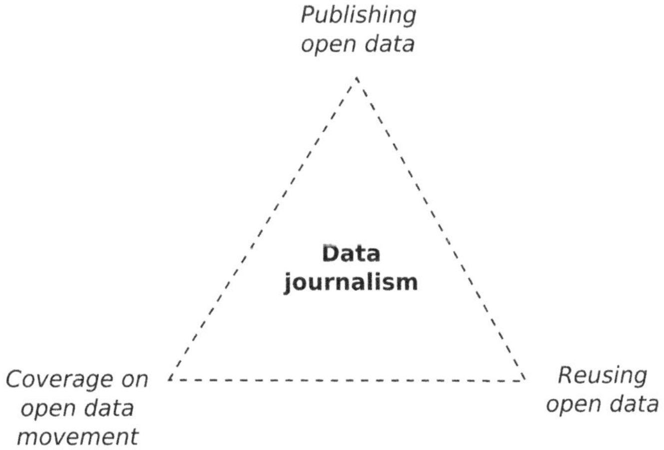

Fig. 6.4 Three pillars of data journalism activities among media actors. *Source: own illustration*

directions and promising genres in journalistic professions, especially in preparing various investigative journalistic materials. Along with the widespread popularity of centuries-old traditions of freedom of information, the modern concept of open data ideally fits to the growing popularity of the open knowledge movement in these two Nordic nations that propagate the ideals of open information society. In the favorable context of social democracy and political representation in advancing e-government policies, these values are actively supported and promoted by the public sector. Open data is widely regarded as one of the effective instruments to achieve the goal since any data-driven initiatives require greater collaboration and networking among various stakeholders for their successful implementation, including media communities.

2 Theoretical Contributions of the Research

The primary theoretical contribution of the work is the presentation and discussion of new hypotheses, assumptions and theories that were generated in the course of multiple case study research and further analysis of its key results and evidence derived from empirical observations and a myriad of cases in the context of two Nordic nations. In this regard, the close investigation of open data policies adopted by national and local authorities in Finland and Sweden as well as the analysis of promising networking trends in the open data movement could enrich available knowledge about the dynamic development of this new phenomenon in its real-life contexts and, more importantly, understand better the implications of political and socioeconomic character, especially in analyzing typical drivers and barriers that different stakeholders both public and private face in adopting various open data-driven collaborative and civic engagement platforms and related startups in the area. The main hypothesis of the research suggests that the active promotion of the concept in private and non-governmental sectors leads to the creation of new market niches where a promising phenomenon of open data-driven entrepreneurial and philanthropic activity of middlemen and intermediaries emerges, especially among a wide community of independent developers, technically savvy citizens, journalists, academia, business people and e-government enthusiasts, which, regardless of different socioeconomic contexts presented in the research, could dramatically change traditional channels of public sector reforms in any country, theoretically making it more sustainable as a political and socioeconomic phenomenon due to an increasingly higher level of public-private partnership in the area, transform traditional philosophy of the top-down realization in open data governance and freedom of information regulation with new collaborative and networking dimensions and open new opportunities to boost public sector innovations not only in e-government related areas but also in the private sector, too. Moreover, the author of the book also presents new hypotheses on implications of such new administrative trends as digital centralization and decentralization that could be observed today in open data regulation strategies and related bureaucratic practices in many countries, too, changing traditional understanding of how government datasets are generated, stored and released in different institutional environments.

Another theoretical contribution of the work is the presentation of its more universal research framework for multiple cross-country investigations in the area, which could be helpful in studying open data governance also in other countries in an intrinsically consistent manner, enriching existing methodological models that are usually utilized in related comparative case studies. In this regard, such aspects of the analysis presented in the work as, for example, *a scattergun approach* that includes the investigation of the open data phenomenon from a wide range of perspectives such as *a country context analysis* that studies the implications of national history, political and social systems, economy, public administration, telecommunication industry, non-governmental sector, mass media and even foreign policy on the development of the phenomena in each of the nations; *a content analysis* that investigates the key open data-driven projects realized at national and local levels; and, finally, *a policy analysis* that speculates on the apparitional emergence of open data-driven networking and e-participation trends in the area could help researchers in investigating the development of an open data phenomenon in other countries in all their diversity. The intrinsically multidimensional approach in the methodology of the case study research aimed at examining as many political, technological and socioeconomic domains of the phenomena as possible proved to be consistently effective in receiving rich empirical data from all aspects of the open data movement in all cases that have been observed in consecutive order in these two nations, which, in turn, could be replicated in other comparative studies in the area.

3 Practical Contributions of the Research

The main practical contribution of the book is that it provides a detailed review of open data governance from the perspective of various stakeholders who are usually engaged in the promotion of open data-driven projects that could be especially helpful for policymakers and practitioners in evaluating and improving the work of various innovative startups in the area. One more practical contribution of the research is a number of policy recommendations that the author presents in his work for open data and e-government practitioners, project managers and operators in solving various challenges and issues in a wide range of open data platforms and related digital initiatives realized in these countries at national and local levels of governance. In this regard, Chaps. 4 and 5 provide a concrete set of advices on how to improve administratively, economically and politically the operation of various open data and, more importantly, open data-driven civic initiatives such as on how to: a) promote universal and more sustainable models in realizing related third-party projects; b) boost civic engagement and independent fundraising trends in the area; c) grasp the attention of business communities and developers and motivate them to join the movement and contribute more actively; d) reduce the burden of the hidden economic cost in open publication of government datasets that some semi-public agencies could deal with; e) raise the public awareness in the area with the help of journalistic communities and social media campaigns; f) develop socioeconomic models of mutually beneficial cooperation between open data entrepreneurs, independent developers and non-governmental organizations that would result in better quality of government datasets, provision of completely new open data-driven services within the traditional ecosystem of e-government and, more importantly, open new market niches for potential investors in the area. The author also indicates the potential emergence of new challenges that may impede the development of the concept and offers a set of practical recommendations on how to overcome them, too.

In general, the results of the research could be used by practitioners and policymakers in evaluating and improving the operation of various open data-driven projects all over the world, providing a unique multidimensional and cross-border perspective on the development of this promising phenomenon, especially taking into account that it is analyzed in the context of two developed economies that currently lead the open data movement globally. In this regard, the author of the book claims that it is both a policy review and an agenda-setting research, which is intended for a global community of open data experts, e-government specialists, political scientists, journalists, lawyers, developers, business persons and all those who are interested in studying this promising political and socioeconomic phenomenon. Open data governance is a completely new phenomenon, which could affect many aspects of human activity and change the traditional understanding of technology-driven public sector reforms in many countries. The book, which presents many ideas in a unique graphically illustrative manner, is mostly intended as an introduction to this emerging concept from both theoretical and practical perspectives with many case studies from different stakeholders and from different political and socioeconomic contexts and will be ideal for various practical trainings and academic courses in the area.

4 Limitations of the Research

4.1 The Uniqueness of the Cases

Probably, one of the fundamental limitations of the case study is that its findings and hypothesis could relate only to a limited number of countries that have similar political, socioeconomic and cultural contexts, decreasing the potential reading audience of open data and e-government researchers and practitioners who may be generally interested in studying the development of open data governance in many developing and emerging nations. In this respect, cultural contexts may play an important role in the development of unique political and socioeconomic conditions favorable for the promotion of open data as a phenomenon. It is especially true when it comes to such developed social democracies as Finland and Sweden. The widespread popularity of extensive social security and social welfare programs in these and all other Nordic nations (e.g. Iceland, Denmark and Norway) has created really favorable socioeconomic conditions for extremely developed and, more importantly, self-sufficient communities of independent developers who are ready to contribute their time and personal resources and help their local neighborhoods with their innovative open data-driven solutions and related digital startups, often on a philanthropic basis, driven mostly by professional curiosity or feelings of social responsibility. In addition, these Nordic countries have historically been true leaders in advancing open knowledge phenomena even before the advent of the digital era, which makes them all unique from historical, socioeconomic and political perspectives. In this regard, in order to mitigate partly the objective limitations in investigating such a specialized and, more importantly, context-dependable topic, the author of the book has also applied a scattergun approach in analyzing open data with the main focus on the combinative use of context, content, network and policy analysis that might be helpful in investigating it as a multidimensional phenomenon in an attempt to provide a universal conceptual framework for related case study research, regardless of the context, thus, increasing its potential readership from other regions of the planet, too.

4.2 *Primary Focus on National Open Data Projects*

Another limitation of the case study is that the author of the research tries to analyze the development of open data governance only at national levels, paying less attention to the implementation of local projects in the area. One of the primary reasons why the author did not study the local projects is that most of them could not be considered politically independent since operationally they are all based on the same principle of digital centralism that is widely applied in these unitary countries, which is reflected in the use of single databases and unified processing centers, making them, to a greater extent, regional representations or branches of the national open data project at local levels with only insignificant differences in the organizational adaptation and realization of the platforms in accordance with local needs and funding opportunities. The second reason is that despite the fact that the policy analysis and the field study of the related projects at local levels could be really interesting in understanding the implications of regional administrative, socioeconomic and even cultural contexts on the development of open data phenomena, it would have been an extremely time- and a resource-consuming undertaking that was beyond the capability of a single researcher within this case study.

4.3 *Limitations of Statistical Resources*

One more objective limitation of the research is that the investigation relies heavily on the analysis of statistical data from open depositaries which are mostly operated by national statistical agencies and central government bodies that could raise questions in the academic community about the potential biases of the information provided by only public agencies. In this regard, in order to mitigate partly the possible effects of the restraint, the author of the research has also resorted to the analysis of statistics provided by global intergovernmental agencies and non-profit organizations. Therefore, methodologically, the case study is a qualitative research, which is based on the analysis of primary data received during various semi-structured interviews and meetings with various stakeholders and use of secondary statistical data. The main aim of the research was to study open data governance as a multidimensional phenomenon, analyzing, first of all, political, socioeconomic and cultural contexts existing in these two societies and its indirect implications on the development of the concept from both theoretical and practical aspects and from a variety of perspectives, both stakeholder and institutional ones. In this respect, the research claims to be an agenda-setting and policy review study, the findings of which could be generally transferable and extrapolated to other cultural settings, regardless of the context.

Glossary

Citizens-sourcing is a self-sustainable voluntary mechanism in generating and providing new information, data, services or ideas by local neighborhoods for a particular communal project or initiative.

Civic engagement is a policy that is aimed at engaging a wide community of citizens to join a particular collective project or movement, which could result in the creation of public good or public values, contributing to its overall development and progress with their knowledge or finances.

Crowdfunding is a philanthropic movement that aims at voluntary generation of financial and material resources among citizens, which could be necessary to solve one particular communal challenge or public issue.

Developers are a group of experts in information and communication technologies, who are professionally engaged in creating and running various digital projects and startups in the area.

Digital democracy is a political concept that propagates the use of new information and communication technologies to increase the level of political participation from the part of local communities and public sector reforms, especially in relation to the cardinal transformation of communication channels that connect government and citizens.

M. Kassen, *Open Data Governance and Its Actors*, Studies in National Governance and Emerging Technologies,
https://doi.org/10.1007/978-3-030-92065-4

Digital state is a ubiquitous penetration of various digital and technological solutions in a wide range of spheres both public and private in one particular nation at both national and local levels of governance.

E-centralism is a special ecosystem of e-government that promotes the development of centralized implementation policies in the area at both national and local levels, ideally with the creation of single databases, single digital government platforms and centralized identification of citizens.

E-confederalism is a special ecosystem of e-government that promotes the development of multidimensional, cross-border and multi-institutional implementation policies in the area at regional, national, provincial and local levels, ideally with the creation of a wide range of cross-border databases, multiple digital government platforms and decentralized identification of users at various regional levels of e-governance (e.g. in the European Union).

E-federalism is a special ecosystem of e-government that promotes the development of different implementation policies in the area at federal, provincial and local instances, ideally with the creation of various autonomous databases, multiple digital government platforms and decentralized identification of citizens at national and local levels of e-governance.

E-government is a combination of innovative policies, technological platforms and digital services that are used in public administration and public sector reforms.

E-participation is a political concept that intrinsically propagates the increased level of collaboration between government institutions and other stakeholders, for example, citizens and businesses, in decision-making processes in order to improve the work of public agencies or accelerate the implementation of public sector reforms.

E-voting is the use of various information and communication technologies in organizing and ensuring secure and secret voting or calculating ballots.

Hackathon is a special online event or actual physical meeting, which is aimed at networking various developers, designers, coders and other professionals in information and communication technologies, brainstorming innovative ideas, developing and promoting new startups in e-government and e-commerce areas.

Independent developers are a group of experts in information and communication technologies or a community of technically savvy citizens, who are directly engaged in initiating, supporting and running their own digital projects and startups in the e-government area on a philanthropic basis, driven mostly by a mere desire to help local communities with their digital solutions.

Mashup is a special hybrid web-application that aggregates, reuses and processes different types of datasets from various sources both public and private, which could be used, for example, to generate new ideas and offer new proposals on how to improve digital government systems and services.

Open data journalism is an active use of various public and semi-public datasets by journalistic communities in preparing their own media products, investigations and digital solutions in the area.

Open data platforms are public or semi-public digital depositaries that publish various types of government datasets in machine-readable formats.

Open data is also known as open government data and is a combination of various formats and types of government information and datasets, such as statistical and contact information, archive recordings, geostationary and geographic data, video, audio and photo files and so on, that are publicly available in special government depositories and, more importantly, in machine-readable formats.

Open data-driven projects are third-party startups that are usually initiated and run by technically savvy citizens, independent developers and open data entrepreneurs themselves, using various types of government data as a raw material.

Open data governance is a management of government policies aimed at disclosing various types of public datasets in a cross-institutional manner, which could be later used by a wide range of private and civic stakeholders to create new independent e-government products and services.

Open government is a political concept that intrinsically propagates the values of transparency, collaboration and participation in government institutions and related political or public sector reforms.

Open knowledge movement is a global philanthropic movement, which explicitly propagates the public value of free access to information in its various forms.

Open science is a philanthropic concept popular in academic communities, which proponents defend the ultimate right of researchers, regardless of their incomes and status, to freely access any scientific documents from special public depositories and digital libraries without any individual or institutional subscriptions.

Open services are a set of open data-driven services that could help developers and entrepreneurs to apply already used and successfully tested algorithms and computer codes in the area, ideally on a free-of-charge basis.

Open society is a political concept, which propagates fundamental values of civil society and individual citizens to be actively engaged in local politics, freely communicate with peers and government and express their concerns in the most open and proactive manner.

Open source software is a type of computer programs and mobile applications that is developed by using open technological codes, which could be then freely readable, modified and duplicated in any computers and other digital systems, usually without any copyright restrictions and often on a free-of-charge basis.

Pirate politics is a combination of policies aimed at propagating the fundamental rights of citizens to have an anonymous, neutral and unlimited access to any digital content without any copyright restrictions and government regulation.

Public sector innovations are technological achievements that result in the significant improvement in the operation of government institutions or implementation of public sector reforms, for example, reflected in the increased generation of new electronic services, cost-effective digital solutions in public administration, e-government, e-voting and e-participation.

Public value is a main political and socioeconomic benefit that is achieved as the result of realizing a particular concept, carrying out public sector reforms or implementing innovative projects.

Public-private partnership is the close cooperation between various government agencies and business entities in implementing or supporting public sector reforms or launching innovative projects.

Social democracy is a political concept or ideology that explicitly promotes the values of social justice and equality in managing economic and public reforms in a wide range of areas (e.g. social security, education, healthcare and taxation) with an ultimate goal to ensure widespread prosperity and individual freedoms.

Startup is a collaborative civic or business project initiated by independent developers and technically savvy citizens to boost innovations and networking.

Technically savvy citizens are a group of citizens who are interested in the latest developments in information and communication technologies and have the knowledge and skills at levels that are sufficient to initiate, create or support their own digital projects and startups in the area.

Technological half-life of open data is an initial active period of innovations and development in many open data-driven startups, which could overtime degrade or be left abandoned due to lack of socioeconomic motivations and opportunities for owners and independent developers to continue these projects in a sustainable and cost-effective manner.

Technology-driven public sector reforms are a combination of policies and strategies aimed at boosting public sector innovations through the use of new information and communication technologies.

Transparency of government is a political concept that propagates the values of openness in government and its readiness to cooperate with private and non-governmental sectors of the economy, especially in implementing various public sector reforms and publishing information and reports on its various activities.

Wiki-based project is collaborative digital platform or startup driven by the power of civic engagement and citizen-sourcing in generating a wide range of additional information and data, which are necessary to aggregate in a voluntary and, sometimes, anonymous way for its normal functioning.

Index

M. Kassen, *Open Data Governance and Its Actors*, Studies in National Governance and Emerging Technologies,
https://doi.org/10.1007/978-3-030-92065-4